The Anarchist in the Library

The Anarchist in the Library

How the Clash

Between Freedom and Control

Is Hacking the Real World

and Crashing the System

To Robert,

SIVA VAIDHYANATHAN

Thanks for reading this.

Siva

BASIC
BOOKS

Basic Books
A Member of the Perseus Books Group
New York

Published by Basic Books,
A Member of the Perseus Books Group

Paperback edition published in 2005
Hardcover edition published in 2004

Books published by Basic Books are available at special discounts for bulk
purchases in the United States by corporations, institutions, and other
organizations. For more information, please contact the Special Markets
Department at the Perseus Books Group, 11 Cambridge Center, Cambridge
MA 02142, or call (617) 252-5298, (800) 255-1514 or e-mail
special.markets@perseusbooks.com.

Library of Congress Cataloging-in-Publication Data
Vaidhyanathan, Siva.
    The anarchist in the library : how the clash between freedom and control
is hacking the real world and crashing the system / Siva Vaidhyanathan.
      p.   cm.
    Includes bibliographical references and index.
    ISBN 0-465-08984-4 (hardcover)
    1. Information technology—Social aspects. 2. Information technology—
Political aspects. 3. Internet—Social aspects. 4. Internet—Political aspects.
5. Information society. I. Title.
T58.5.V35 2004
303.48'33—dc22
                                                                    2003026089

ISBN 0-465-08985-2 (paperback)

Designed by Brent Wilcox

For Melissa

I am a citizen of the Kosmos, the Universe.
Diogenes of Sinope

I am a citizen of the world—I'm a drunkard.
Rick of Casablanca

CONTENTS

INTRODUCTION

One hot, peaceful day in July 2001, the front page of my daily newspaper showed a picture of Russian President Vladimir Putin smiling and embracing Chinese President Jiang Zemin. They had just signed a treaty of "friendship and cooperation" spurred by growing conflict with the United States. Both governments had received stinging criticism for cracking down on what they considered troublesome speech. China came under pressure to release American scholars arrested while doing research in China, and Russian officials were accused of stifling internal dissent after Putin took office. In voicing concerns about the treatment of dissenting voices and academic researchers, the United States took the moral high ground.

So it was somewhat incongruous when, later that same day, U.S. officials arrested a Russian computer scientist and charged him with violating an American law while he was working in Moscow: writing the wrong computer program.

The FBI detained the Russian software engineer, Dmitry Sklyarov, after he gave a speech about encryption at Defcon, an annual hacker convention in Las Vegas. Sklyarov, a programmer for the Moscow-based company Elcomsoft, was about to board a plane at the Las Vegas airport when federal agents accosted him. A warrant had been issued the previous month by the Federal District Court for Northern California in San Francisco charging Sklyarov with a criminal violation of the Digital Millennium Copyright Act (DMCA) for revealing weaknesses in an encryption scheme that the software company, Adobe, employed in its electronic book-reading program. His speech over the weekend outlined how weak and easy to crack the Adobe system was.

Sklyarov was not arrested for talking about decryption but for composing the software that unlocked the Adobe eBook Reader encryption system. His software turned protected digital eBook files into the easily accessible portable document format (PDF), which makes them readable on more than one computer and easier for sight-impaired people to use. Until the summer of 2001, Sklyarov's employer, Elcomsoft, sold copies of the decryption software, called Advanced eBook Processor, for $99. Adobe, like so many other firms that trade in culture and information, was worried that its products could be shared without remuneration by millions of users of peer-to-peer communication networks. After all, in 2001 U.S. courts ruled that Napster, the company that sponsored the most notorious music file-sharing system, had contributed to the copyright infringement of millions of songs.

Dmitry Sklyarov's arrest caused a minor international uproar. Computer scientists and encryption experts announced they would not travel to the United States, since their work might be misinterpreted as a threat to American businesses. "Free Dmitry" appeared on Web sites and T-shirts. Adobe was shamed into announcing that it was not behind the move to prosecute the scientist. The federal government eventually dropped the charges against Sklyarov in exchange for his agreement to testify against Elcomsoft. In the fall of 2002, a jury found Elcomsoft not guilty, and jurors wondered how the DMCA could be enforced against a foreign national or company for work done outside U.S. borders.

The Sklyarov case raises some important questions about how the information environment will be regulated in the twenty-first century. Can a cultural industry survive or thrive without high protection from the state? Should governments try to enforce information policies across borders and oceans? Is it possible to prevent the invention and distribution of a small batch of code or an algorithm? What implications do such laws and technologies have for imagination, expression, adaptation, and aggregation of culture and information? What impact will these laws have on science and mathematics research? What does it mean for the future of democracy when a nation prosecutes someone for opening up the electronic text of Aldous Huxley's *Brave New World*?

The Sklyarov case shows how quickly battles to control digital communication can spill over into the "real world." The issues of democracy, sovereignty, and control in the digital world are not new. Until now the debate has focused on whether we should and how we could install "friction" into an otherwise unregulated medium; about how closely we should try to make cyberspace conform to and resemble the analog world. The metaphors we use to discuss controls in cyberspace always appear clumsily lifted from our more familiar transactions: locks, gates, firewalls, crowbars, vandals, and shoplifters. Those with a vested interest into the status quo have been trying to lock their economic and communicative advantages into the new electronic networks. Others have been arguing that cyberspace is so fundamentally different that no rules should apply. What neither of these sides has acknowledged, what makes these issues so urgent, is that instead of cyberspace becoming more like the real world, the real world is becoming more like cyberspace. It's become trite to say we live in an information world in which bytes mean as much as atoms. But lately the debate over topics as urgent and diverse as cultural identity and national security suggests that we have allowed our virtual worlds to invade our real worlds, altering forever the terms we use to describe our problems and the tools we use to manage them.

Oligarchy Versus Anarchy

This is a story of clashing ideologies and dizzying technologies. The ideologies did not arise with the popularity of America Online or the merger of Vivendi and Universal. In fact, they are among the oldest ideologies still around: anarchy and oligarchy. Anarchy is a governing system that eschews authority. Oligarchy governs from, through, and for authorities. These ideologies feed off each other dialectically. Oligarchy justifies itself through "moral panics" over the potential effects of perceived or imagined anarchy. Anarchy justifies itself by reacting to alarming trends toward oligarchy. Anarchy and oligarchy have new resonance in our digital, connected age. These ideologies are rapidly remaking our global information ecosystem, and the information ecosystem is remaking these ideologies. Those of us who are uncomfortable with either vision grow increasingly frustrated with the ways our media, cultural, information, and political systems are changing. We thought we had gone a long way toward disposing of anarchy and oligarchy, but they are back in slick new forms.

Freedom can be terrifying. Cultural and technological trends are increasing freedom in ways many people find threatening. Yet the reactions (or more accurately "preactions") to these trends are extreme, ill-considered, and imposed unilaterally without public discussion or deliberation: easy answers to difficult problems. More often than not, we have used technological quick fixes to avoid complex, serious discussion of the dangers posed by the increasing speed and amount of information. I hope to prompt more careful thinking about how much and which freedoms are excessive or dangerous. And I hope to identify and criticize "moral panics" engendered by the common perception that freedoms are getting out of hand, that the anarchists are taking over the libraries.[1]

More centrally, I am concerned about the blowback against the specter of information anarchy, and the ways those reactions constitute threats to the widely accepted freedoms to browse, use, reuse, alter, play with, distribute, share, and discuss information. These are valuable behaviors that help creators and citizens shape their worlds. The strange interactions among technologies, ideologies, and desires that have emerged in the past decade have opened up new ways to be creators, consumers, and citizens.

This book offers an account of the information arms race. One side invents a device, method, algorithm, or law that moves our information ecosystem toward increased freedom of distribution and the other subsequently deploys a method to force information back into its toothpaste tube. This pattern imposes the basic paradox of the digital world onto the real world: Stronger efforts toward control often backfire to create less controllable—and less desirable—conditions.

Techno-Fundamentalism

Two revolutionary technological phenomena—digitization and networking—have amplified the sociopolitical trends we call globalization to stir up all kinds of messes. The confluence of these dynamics challenges and complicates things we used to take for granted: the stability of nation-states; the sense that we could afford to embrace certain freedoms because most people couldn't actually exercise them; and the unacknowledged comfort that distance and inconvenience offered us in an earlier, less connected world. Inconvenience was comforting: Potentially dangerous information and alarming expressions only circulated in obscure pockets of subculture or lay undisturbed in inaccessible repositories. The bad stuff was always around. It was just inconvenient to find, distribute, or deploy.

The collapse of inconvenience has sparked a series of efforts that could reestablish the distance, or friction, that our information systems have featured since the rise of movable type and bound books. Often the imposition or creation of friction involves building a new machine to correct for the last machine's consequences. This blind faith in technology as a simple solution to complex social and cultural issues is what I call "techno-fundamentalism." It has ideologically colored the discussions of public concerns ranging from pornography to piracy to national security.

Those who seek to restrict the flow of information use two rhetorical strategies when campaigning for the techno-fundamentalist changes that would empower them. First, they argue for treating information as property or contraband. An appeal to "property" removes information policy discussions from the domain of the public interest. And an appeal to "contraband" nudges the public to surrender freedom for the sake of imagined security. But there is a special risk in regulating information and the technology that delivers and processes it, a risk absent from efforts to control objects or devices such as cars or guns. Information is special. It is the raw material of deliberation. And rich deliberation is the foundation of healthy democracy.

Because these battles have been waged with blunt technological tools instead of intricate, sophisticated, and messy human deliberations, there has been no systematic examination of the long-term effects that such technological regulation might generate. More often than not, deploying a simple technological tool to confront a complex social or cultural phenomenon causes more harm than good. The same tools used to restrict music sharing may be used by oppressive states to crack down on political dissidents. Efforts to stifle communication among terrorists may undermine legitimate uses of networks and information by students or scientists.

This book takes as its initial subject the battle for control of peer-to-peer communication systems, such as Napster, Gnutella, and the Internet itself. But the future of entertainment is only a small part of the story. In many areas of communication, social relations, cultural regulation, and political activity, peer-to-peer models of communication have grown in influence and altered the terms of exchange of essential information and cultural forms. In recent years several broad global systems have developed in which information flows freely through unhierarchical channels. This book examines how traditional power structures such as corporations and governments have reacted to the rise of these systems. The terms and tactics of freedom and control—so familiar to those who follow the debates over the effects of the Internet and efforts to regulate it—are now spilling over into battles that will shape what we mean by culture, liberty, democracy, and human progress in this new century. This book thus considers the ways that similar strategies for stifling electronic communicative networks support efforts to rein in such "real world" systems as Falun Gong, open source programming communities, scientific cultures, anti-neoliberal protest movements, dissident political groups, globalizing religions, terrorist organizations, and systems for counting votes in the elections of the twenty-first century.

As distributed information systems gain prominence and importance, the reaction to them grows fierce. Distributed systems tend toward anarchy. Centralized systems tend toward oligarchy. The space between these models is shrinking, offering no middle ground, no third way. Efforts to minimize the effects of too much freedom or too much information-based contraband tend to quash legitimate uses, as well as the flow of beneficial content. Because both the anarchists and the oligarchs prefer to use technology first and ask questions later, they are avoiding the hard work and the inevitable compromises that an open, humane, connected, stable, liberal society requires. When we see a need to curb freedoms, we should do so deliberately, soberly, and carefully. Yet in the early days of the twenty-first century threats to freedom are arbitrary, hasty, panicked, and often brutal. One of the chief challenges of the twenty-first century will be to formulate ethics, guidelines, habits, or rules to shape an information environment that provides the freedom liberal democracy needs as well as the stability that commerce and community demand.

Anarchy and oligarchy are rapidly remaking our information ecosystem and challenging the way we talk about commerce, globalization, and citizenship. Those who are uncomfortable with both ideological visions and value republican "information justice" increasingly find fault with changes in media, cultural, information, and political systems. There are many examples of conflict arising out of efforts to control the flow of information:

- The "locust man," an imprisoned democratic activist in China, distributed political messages by attaching them to the backs of locusts.

- The ordeal to which public libraries in Fairfax County, Virginia, and Broward County, Florida, were subjected when some of the September 11 hijackers used public terminals in the days preceding their attack. Under the provisions of the USA Patriot Act, an increasing number of American librarians have felt pressured by federal law enforcement agencies that might ask them to violate their code of ethics and their patrons' privacy since this incident.
- The controversy over the complaint that some Canadian women cannot be tested for genes that indicate a predisposition for breast cancer because an American company has patented those genes and charges too much for the test.
- The widespread acts of civil disobedience that have led otherwise liberal institutions like Swarthmore College to bow to corporate pressure and shut off Internet services to students who shared damning and revealing information about a major vendor of electronic voting machines.

These and other case studies raise a host of important issues:

- The battle to control public libraries, which are suddenly considered dens of terrorism and pornography, through technological mandates and legal restrictions.
- Efforts by governments to radically reengineer personal computers and networks to eliminate the very power and adaptability that makes these machines valuable so that they can better control flows of material deemed illicit: pornography, copyrighted music, or raw materials of political dissent.
- The cultural implications of allowing fans and creators worldwide to sample and share cultural products at no marginal costs through peer-to-peer computer networks.
- Attempts to restrict the use and distribution of powerful encryption technology out of fear that criminals and terrorists will evade surveillance.
- Commercial and governmental efforts to regulate science and mathematics, including control over the human genome.
- Attempts to stifle the activities of political dissidents and religious groups.

- Proposals by the U.S. government that constitute domestic "information warfare," such as the USA Patriot Act, the aborted TIPS initiative, and the Total Information Awareness program.

Peer-to-peer systems are about more than music. Individuals are disorganizing to distribute an abundant resource that the state and other powerful agents wish to make scarce. The battles over control of cultural distribution can be read as a prelude to more overtly political battles to come. The peer-to-peer model of information sharing is both ancient and emergent. Peer-to-peer electronic networks simulate transactions that have occurred throughout human history in open, nonhierarchical systems. Gossip is anarchistic and peer-to-peer. Communities of punk rock fans sharing home-replicated cassette tapes are peer-to-peer. So are most major world religions. But something has clearly changed. Global electronic, unmediated communication, often masked by encryption, has collapsed time and space. Global communication and organization by nonelites is stimulating a reaction by elites. Powerful institutions cannot ignore the threats that open, anarchistic systems seem to present. Thus reactions to the emergence of peer-to-peer systems—though largely unjustified—have been fierce.

Taking Anarchy Seriously

Anarchistic structures and tactics matter to our daily lives more each day. Implicitly, therefore, important social, spiritual, and communicative networks are building themselves along the principles of radical democracy—or, as it is sometimes called, anarchy.

Whether we embrace anarchy or fear it, we should try to understand it. Anarchy matters. Peer-to-peer systems, like other distributed systems, are like punk rock: They empower fans and citizens, create new communities, and close the gap between creators and consumers. They democratize elements of cultural production and demand a new set of theories. But anarchy is not common democracy, at least not in the sense that we have been accustomed to debating and fighting for through the twentieth century. Democracy usually requires stable procedures and protection for minorities against the tyranny of the majority. It has in recent times been tempered by republicanism.

Anarchy is radical democracy. It has its limits as a governing tool; it also has its dangers. "Smart mobs" are still mobs. In a mob, anyone who steps out of line or runs at a different pace can get trampled.[2]

The great challenge in the new century is to mediate between two divergent and trends—anarchy and oligarchy. In the war between distribution and concentration of information, the issues and conflicts seem intractable.

I won't accept intractability. We have only begun to consider the long-term ramifications of these revolutionary technologies and the behaviors they will enable or inspire. If we can energize an open, distributed, diverse network of thinkers and writers to consider these conflicts in a new way, using fresh vocabulary and models, we can generate social, cultural, legal, and technical protocols that will strengthen democracy and inspire trust and confidence. If we fail to generate this conversation, if we continue to let these conflicts work their way through courtrooms and technological incubators, basements and boardrooms, both democracy and stability are in danger.

CHAPTER 1

Public Noises

Parisians living in the turbulent eighteenth century found out about their world and their politics by sharing "public noises" *(bruits publics)* in a handful of social nodes around the city. They gathered and gossiped at Pont Neuf, on particular benches in Luxembourg Garden, in various cafés, and most importantly under the Tree of Cracow in the garden of the Palais Royal. Fortunate Parisians would then gather in salons to sift through the gossip and debate the issues of the day, although they weren't supposed to. According to historian Robert Darnton, politics was not supposed to be public. It was the king's business, *le secret du roi.* All printed matter was supposed to be filtered through the king's offices. The illicit discussion of issues, personalities, and scandals was treated as gossip. Gossip was essential to the emergence of the Enlightenment and eventually the French Revolution—and ultimately the bourgeois revolution across Europe.[1]

Gossip is uncensored, unmediated, unfiltered, peer-to-peer communication. The term describes the method, not the subject, of communication. In prerevolutionary France gossip was anarchistic in the sense that it grew and thrived outside of hierarchies and beyond the reach of the state, not that it necessarily advocated or caused the overthrow of the king, although it certainly did contribute. Unmediated and decentralized, gossip was anarchistic in structure if not content. If Parisians were allowed to imagine dissenting from the arbitrary strictures of the sovereign, they might eventually imagine a new kind of a state. Parisians, unlike those who lived outside the city, enjoyed the advantage of geographic proximity and noise that created a shield from the state. The agents of the state lacked the technology, numbers, and will needed to monitor every conversation occurring in every park of Paris. Rural citizens were less likely to experience state surveillance, yet they were excluded from the sites of information exchange. The inconvenience—the expanse of time and space and the expense of traversing them—extended the rule of monarchy for a couple more decades.

Robert Darnton, in examining the ideological origins of the French Revolution (e.g., in *The Great Cat Massacre* and *The Forbidden Best-Sellers of Pre-Revolutionary France*), has described the methods and substance of conversations that escaped state surveillance. Darnton explains how ideological communicative functions informed the revolution that was brewing; in his view, there was more to the French Revolution than the enlightened rationalism of the philosophes. They mattered, but not as much as we might think. Anarchy mattered too.[2]

Taking Anarchy Seriously

Since about 1776, the world has been trying to decide how much centralized control it will allow over its affairs. Models inspired by Adam Smith, Karl Marx, John Stuart Mill, Friedrich Hayek, and John Maynard Keynes have dominated our debates and inspired our wars. As the twenty-first century dawns, we need to reexamine the contributions of Emma Goldman and Johnny Rotten as well as those of Marx and Mill.

Anarchistic systems are a relatively unexamined function of the sociopolitical dynamics of our time. Some of the tangle of changes in the world that we clumsily call globalization includes the establishment of distributed networks of information. Some of the reactions to this process involve efforts to centralize control of those flows under the power of nation-states, multinational corporations, or multilateral international organizations. So we must consider the ways in which these processes challenge the stability and relevance of the nation-state and open a new front in the battle over the value of Enlightenment thinking.

For the past two hundred years, the centralization of power and information under the aegis of the state was considered the greatest challenge to republican forms of government and commerce. But now decentralization and distribution have emerged as the most important political reactions to the expanding power of the global state–corporation partnership that is setting the political and economic agenda for the entire world. For the past two hundred years the world has been consumed by the debate between the socialist vision of economic and political organization and the liberal vision. But in the new century, distributed systems represent emerging anarchic visions, new and unexamined challenges to liberalism's Enlightenment project. Whether we welcome anarchy or fear it, we must take it seriously.

The Ideological Origins of the Techno-Cultural Revolution

Anarchistic functions and methods have been around for many centuries. Recently individuals have used widespread, low-cost, high-quality technologies to communicate, persuade, and organize over long distances, beyond the prying ears and eyes of powerful institutions. Digitization and networking make anarchy relevant in ways it has not been before. Global electronic networks make widespread anarchistic activity possible. What used to happen in a neighborhood barbershop or on a park bench now happens across a nation-state or beyond. Rumors can bubble up into action.

We are in the midst of an unfinished global techno-cultural revolution. Like the French and American Revolutions, it has ideological origins and could have any of several outcomes. It could be a triumph for freedom or unfreedom. It could devolve into violent chaos or a stifling sense of order. Unlike those revolutions, everything is still up for grabs. It remains to be seen which ideologies will triumph and what effects they will have on the world. The present revolution's ideological origins are synergistic and dialectical: anarchism, oligarchy, techno-fundamentalism, and globalism. They are familiar in old contexts and powerful in new ones. Of these anarchism is the ideology that most loudly demands a new look.

Anarchy is not necessarily chaotic and dangerous. It is organization through disorganization—anarchistic tactics generally involve uncoordinated actions toward a coordinated goal. Instead of formal leadership structures, anarchism relies on agreements, conventions, conversation, and consensus to move the group toward action. Anarchy is by definition nonhierarchical, radically democratic. It is possible for nonanarchists to employ anarchistic tactics or join anarchistic organizations in the sense of believing that the state should wither, crumble, or fall. Just as exploiting socialist institutions such as public schools or farm subsidies does not make one a socialist, so protesting the World Trade Organization or using a peer-to-peer service to share music does not entail flying a black banner, wearing a black mask, or smashing a McDonald's restaurant.

Derived from the Greek word *anarchos,* "without authority," anarchism denies law and considers property to be tyranny. Anarchists believe that human corruption results when differences are enforced through the maintenance of property and authority. Anarchists do not oppose or deny governance as long as it exists without coercion and the threat of violence. They oppose and deny the authority of the centralized state and propose governance through collaboration, deliberation, consensus, and common coordination. Justice can emerge from a sense of common purpose and practices of mutual aid, not the monopoly on violence that the state demands. While anarchism is commonly associated with bloody violence and rage, anarchists believe deeply in an ideology of love.[3]

From Public Noises to Public Nuisances

Many are sadly ignorant of the breadth and depth of anarchistic thought and its legacy in world history. In both my undergraduate and graduate curricula, I studied political philosophy as an element of American cultural history. Amid all the Aristotle, Machiavelli, Madison, Jefferson, Marx, and Rawls that I had to read, only one writer took the concept of anarchy seriously enough to mention it: American philosopher Robert Nozick.[4] Anarchy has played a romantic and tragic role in Europe and America. As a fact or condition, one might say that it was the original political philosophy of *Homo sapiens.* The world has witnessed many stateless societies, groups of people who have lived without a dominant authority. As an explicit political philosophy, anarchism dates from the late eighteenth century. Mature, articulated anarchist thought has been an important undercurrent in political history over the past three hundred years. We have been mostly deaf to it because our big battles have been among the forces that oppose anarchy—capitalism, socialism, and fascism.

Anarchy as a cultural and political stance has its roots among the Cynics of Greek antiquity. Its first modern expression, during the English Revolution of 1640 to 1660, came from a group of landless radicals who called themselves Diggers. Drawing on the radical timber of their time, the Diggers advocated love and the abolition of private property. They thought of the earth as a common treasury for all, and so they would walk onto someone's land and start to dig. Finally they were forced to explain themselves to Lord Protector Oliver Cromwell, who was not interested in helping them forge their propertyless utopia in Albion.[5]

Anarchist thought matured during the eighteenth and nineteenth centuries through writers such as William Godwin, Pierre-Joseph Proudhon, Peter Kropotkin, Leo Tolstoy, and Mikhail Bakunin. These writers influenced the attempted revolutions of 1848 and the Paris Commune of 1871. Karl Marx's social ontology and theories of history remain essential to anarchist thought, but anarchists—chiefly Bakunin—immediately challenged Marx's faith in the power of a strong state to maximize human happiness.[6]

In the late nineteenth and early twentieth centuries anarchists passionately exerted their limited influence on the processes of industrialization and ethnic urbanization in the United States. The visceral reaction their efforts elicited vastly outweighed their political influence. Anarchists have always been on the extreme margin of American politics, unable to articulate their visions and views beyond the distorting lenses of dominant political ideologies. For a brief period in late nineteenth-century American history, anarchism seemed to matter, but as something to fear or oppose rather than a potential source of political change. Several violent events attributed—perhaps unjustly—to anarchists sparked some important effects. In 1877 anarchists organized a nonviolent demonstration at Haymarket Square in Chicago. They were protesting the deaths of two workers from the McCormick Harvesting Machine Company, who had been killed by Chicago police officers for their union organizing efforts. During the anarchist-led protest, a bomb exploded, killing one officer. In response, police opened fire into a crowd of 3,000 people. Seven policemen died in the ensuing battle, and an unknown number of civilians died at the hands of the police. The riot was followed by a an efficient, ruthless crackdown on suspected anarchists. Courts convicted and sentenced to death eight anarchists for playing a role in the riots, even though no evidence linked any of them to the explosion. They were convicted for espousing ideas that made them accessories to the police deaths.[7]

The industrial anarchist movement in America ended when Leon Czolgosz, a fan of anarchist feminist Emma Goldman, assassinated President William McKinley at the Pan American Exposition in Buffalo, New York, in 1901. Czolgosz was a Michigan-born man of Polish-Prussian descent. He was not a very passionate or informed anarchist. And the anarchists who had crossed his path suspected that he was a police stooge or plant. He was certainly deranged. Seeking greater meaning in the event, reporters and officials immediately labeled Czolgosz an anarchist and—just as significantly—a foreigner. Many people expressed relief that McKinley's murderer was not an "American." Soon after Czolgosz was arrested in Buffalo, Chicago police started rounding up anarchists who might have known him from his brief travels to the city. Emma Goldman was among those detained. Ultimately officials decided that Czolgosz had acted alone and there was no conspiracy to kill the president. The trial proceeded swiftly.[8]

During Czolgosz's incarceration and trial, some western New Yorkers tried to lynch him. This generated widespread protest from African American writers, many of whom explicitly linked the principles of anarchy to the persistent habit of lynching African American men. Booker T. Washington, among others, took the opportunity to remind his fellow Americans that the lynch mob was an anarchistic phenomenon: decentralized, leaderless, antistate. "I want to ask," Washington wrote in the *Montgomery Advertiser* in 1901, "Is Czolgosz alone guilty? Has not the entire nation had a part in this greatest crime of the century? According to a careful record kept by *The Chicago Tribune,* 2,615 persons have been lynched in the past sixteen years . . . we cannot sow disorder and reap order." Washington explicitly denounced what he called the "anarchy of lynching" and called for a proper trial. This sentiment echoed through African American newspapers, many of which were controlled by the Republican party.[9]

For many Americans, McKinley's assassination cemented the stereotype of an anarchist as an eastern European, possibly Jewish, definitely violent revolutionary. Two years after McKinley died and Theodore Roosevelt rose to power, American immigration policy reflected Roosevelt's deep-seated racism and the fear of immigrants and anarchists that he exploited in the days after he assumed the presidency. American immigration laws were sparse before 1903, and they chiefly excluded Chinese immigrants. In 1903 Congress passed a new immigration act that moved restrictions beyond race and national origin to include political thought. Anarchists were now forbidden to enter the United States. The Naturalization Act of 1906 required prospective citizens to swear that they were not "opposed to organized government" and to demonstrate a commitment to social order. Thus members of the radical International Workers of the World could be denied citizenship, and many were. In an effort to rally political momentum and define a common enemy that Americans of many persuasions could unite against, Roosevelt declared a "war on anarchism" and called for the expulsion of "all persons who are known to be believers in anarchistic principles." Many of Roosevelt's reform efforts were intended to stave off the more alarming prospect of radical revolution in America, even though no one could sincerely predict anarchists threatening the foundations of American society, economy, or government.[10]

Persistent fear of anarchists justified the Red Scare of 1919, with its brutal Palmer raids that precipitated widespread imprisonment and expulsion of many suspected of disloyalty to the United States. Among those forcibly exiled was Emma Goldman.[11]

Perhaps the image of fiery anarchists rallying workers against the government of the United States rendered the historical awareness of anarchism shallow and marginal. Anarchists seemed passionate, violent, and unreasonable. Their historical presence was fleeting and abrupt. Consequently, few historians have tried to explain the influence of anarchist thought on American history. In Europe, anarchists were present throughout the evolution of radical political thought in the nineteenth century. But anarchist parties mattered little to the resolution of conflicts and tensions until the early twentieth century.

By the dawn of the Spanish Civil War in 1936, the anarchist movement outside Spain was all but destroyed. The rise of fascist governments in Germany and Italy signaled to anarchists that their movements would not be tolerated in much of Europe. Their great hopes in Russia were dashed by the socialist Bolsheviks in 1917. Although the most influential anarchist thinkers, Mikhail Bakunin and Peter Kropotkin, were Russian, the anarchist movement failed miserably in Russia. During the Russian revolution the socialists destroyed and betrayed the anarchists. Lenin and Trotsky were particularly brutal in their efforts to stamp out anarchism. After the 1917 revolution the small anarchist groups that emerged in St. Petersburg and Moscow found themselves powerless against the Bolsheviks. Kropotkin returned from exile in June 1917 and established a tiny,

ineffective anarchist commune in the village of Dmitrov. Peasant orga-
nizer N. I. Makhno raised an armed force that used guerrilla tactics to
hold a large part of Ukraine until 1921, when it too was crushed. The anarchist movement became extinct in Russia, never to rise again in the
twentieth century.[12]

After World War I and the Russian Revolution, Spain was the only
source of optimism and success for anarchists, who counted such notables
as Pablo Picasso among their fellow travelers. From about 1850 until 1936
the Spanish anarchist movement enjoyed popularity and influence. In
1845 Ramón de la Sagra founded the world's first anarchist journal, *El
Porvenir*. By 1870 Spain was home to about 40,000 anarchist party members and three years later it had about 60,000 members. In 1874 the
monarchy grew alarmed by the popularity of the anarchist party and
forced the Spanish movement underground. But anarchist ideas and
passions continued to circulate as political tensions and working-class
dissatisfaction intensified in the early twentieth century. Efforts to crush
the anarchists only dispersed them, playing into their ideological disposi-
tions and making it harder for authorities to exert direct pressure on the
movement as a whole. This state oppression caused anarchists to change
their strategy from supporting an open, identifiable political party to
forming a loose, distributed network of *sindicatos únicos* that linked craft
and trade unions into local committees. The structure of these *sindicatos*
prevented one person or group from dominating the rest and thus
restrained corruption. The national committee that coordinated the local
committees was elected each year from a different region of Spain to ensure that no individual served more than one term and that no part of
Spain grew accustomed to power. Their management and staffing needs
were fulfilled by volunteers. All delegates were subject to recall by the
members. The *sindicatos* grew from 700,000 members in 1919 to 2 million
during the civil war. A formidable force in Spanish life by the 1930s, anarchists were quick to assert themselves among loyalist factions as events led
to the Spanish Civil War in the 1930s.

Troubled by the rise of fascist leader Francisco Franco, Spanish anarchists sparked several unsuccessful uprisings in the early 1930s. By the
early days of the civil war in 1936, anarchist forces defeated royalist and na-
tionalist troops in Barcelona, Valencia, Catalonia, and Aragon. Early in the

war anarchists controlled much of eastern Spain. Workers took over management of factories and railways in Catalonia. Peasants seized land in Catalonia, Levante, and Andalusia. Many anarchists established agricultural and industrial communes governed by radically democratic principles. As the war dragged on, however, such loose arrangements failed to serve armed struggle. Not surprisingly, anarchists were hard to order around. They were prone to break into conversation at moments when obedience might have served them better. The communist-funded International Brigades were more effective warriors than the anarchists were. Franco's fascists were the best armed, best disciplined, and ultimately most successful warriors in Spain. As the war raged on, even countries that declared firm opposition to fascism failed to support the antifascist forces in Spain. Betrayal, both internal and international, doomed the anarchist movement in Spain.[13]

European anarchism moved from the political to the social and cultural realms after World War II, just as most western European states adopted some form of strong social welfare and liberal democracy as their guiding principles of governance. Socialism allowed some anarchists to exploit state support while undermining state authority. The vestiges of anarchist social passion were visible in Christiania, the squatter section of Copenhagen. It was an abandoned military camp until squatters moved in during the early 1970s. They set up a "free state" and governed collectively. Residents still live rent-free and tax-free and run their own schools and parks in Christiania.[14] Anarchists, calling themselves "provos," played a central role in urban reform debates in the 1960s and 1970s in Amsterdam.[15] The 1968 revolts in the streets of Paris saw passionate activity by all elements of the anticapitalist and antiestablishment left. Anarchists were central to the protests, though irrelevant to the aftermath. Every major city in Europe now has an anarchist collective of some sort. For a while postwar anarchists embraced absurdity, play, hedonism, and spontaneity instead of concerted political action. Since the rise of the European Union and the World Trade Organization, European anarchist activism has become more confrontational and politically effective. Thanks to widespread access to global electronic networks across western Europe, activists communicate anarchistically and swarm spontaneously to disrupt any function in Europe deemed worthy of their disdain.

Don't Follow Leaders,
Watch the Parking Meters

The American civil rights movement protests of the 1950s and 1960s had the peculiar effect of both making anarchist tactics relevant and disguising them as Christian. By embracing direct action and civil disobedience, American civil rights leaders did more than advance the centuries-long struggle for racial justice in the United States. The battles waged in courtrooms and legislatures legitimized the anarchistic tactics that protesters were employing in streets and public facilities instead of the anarchistic tactics delegitimizing more republican forms of activism. The civil rights movement is the strongest example of nonanarchistic activists borrowing moves from anarchists for use toward nonanarchistic ends. Cultural, social, and political dissatisfaction with what C. Wright Mills called the "power elite" drove American experiments in anarchy from the focused agenda of the traditional civil rights movement to a broader variety of struggles and efforts through the 1960s. Soon the antiwar, feminist, Asian American, and environmental movements were dabbling in anarchistic forms of governance and civil disobedience. In the late 1960s, a San Francisco collective known as the Diggers—named after the Diggers of the English Revolution—revived anarchism as a cultural, albeit politically impotent, force. They were all about the political power of cultural play. They were devoted to living according to personal maxims and disturbing those who in their judgment were not. According to journalist Don McNeill, who chronicled the 1960s Diggers in his 1970 book *Moving Through Here,* "the diggers declared war on conditioned responses. They blew minds by breaking subtle mores. They practiced public nuisance." The Diggers held playful, silly "happenings" meant to disrupt a structured event. They spread chaos and confusion by spoofing, singing, and screaming, creating a kind of "guerrilla theater." In one incident, they mocked and flaunted a San Francisco law against being a "public nuisance" by walking down Haight Street wearing animal masks, carrying a coffin as if they were pallbearers. Believing in freedom of drugs as well as freedom of speech, they gave away marijuana. They ran a "free store" that offered "liberated" goods at no cost. As Digger Peter Berg explained to Todd Gitlin, the mission of Digger anarchism was to "create the condition that it describes."[16]

But playful anarchism did not change much in the world. Frustration with power structures demanded stronger action than creating a nuisance. Some radical students in the United States, Mexico, Japan, and France began sharing new politics and new possibilities. Being young and bold, these students found the rigid structures of the "old left" dull and ineffective. The soul of anarchism—spontaneity, theoretical flexibility, simplicity, local autonomy, and hedonism—appealed strongly to these young people. In the mid-1970s, anarchistic political energy in America dissipated, but its cultural significance continued to grow. Anarchist theory influenced the rise of punk rock in the United States and England, and anarchist practice influenced the rise of hip-hop culture. By the end of the twentieth century, anarchy was youth culture.

Politically, only the radical ecology movement openly embraced anarchism. Decentralized organizations such as Earth First! took often risky action to protect resources such as old-growth forests. By the 1980s many environmentalists came to believe that the assumptions about the nature of "progress" embedded in both republican and capitalist thought were fundamentally incompatible with efforts to preserve natural resources and species. So instead of engaging in traditional activism—fund-raising, holding press conferences, lobbying, and compromising—the most radical members of the environmental movement resorted to "direct action." Frequently it was nonviolent, such as protesters chaining themselves to a bulldozer or sitting in a tree and daring loggers to cut it down. But some went farther; for example, members of the Earth Liberation Front drove metal spikes into trees that could kill unsuspecting loggers. And more recently, environmental radicals have been destroying dealerships that sell sport utility vehicles in an attempt to drive up the insurance costs of such dealerships and thus shrink the profit margin of their operations. The Earth Liberation Front, like some animal rights activists and white supremacists, practices "leaderless resistance," a political structure meant to simulate a network in which no one member knows everything that the group is up to. If leaderless resistance works (and there is doubt about its efficacy), then every activist or terrorist has "deniability" and thus cannot be held responsible for the actions of others. More importantly, no defection or leak would bring down the entire group.[17]

The environmental critique of "progress" echoes radical political thought of the past three decades. Similar ideas about the pernicious effects of "totalizing" ideologies such as capitalism and Marxism have influenced critical thinkers in the domains of culture, imperialism, and science. Two very different writers have influenced recent thought around the world: Michel Foucault and Noam Chomsky. Foucault, a French philosopher who rose to prominence during the Paris revolts of 1968, did not openly embrace anarchy as his political philosophy. But his quasi-historical work on prisons, asylums, sexuality, and ideology in general revealed the almost invisible forces limiting our otherwise radical sense of freedom and autonomy. Foucault has inspired many cultural and political anarchists, as well as nonanarchists, to experiment with radically libertarian notions. Chomsky, who differed with Foucault in many ways (e.g., on the plasticity of human nature), also rose to fame in the 1960s. Chomsky is a major linguistic theorist at the Massachusetts Institute of Technology who stirringly rebuked American aggression in Southeast Asia long before the rest of the country noticed that it was at war. Chomsky has issued rants and tirades—some verging on paranoia—about various injustices perpetrated by the U.S. government and its partners. Calling his political philosophy "libertarian socialism," Chomsky bluntly endorses anarchistic means to anarchistic ends. He has held up the Israeli kibbutz system—at least in its early incarnation—as a successful experiment in libertarian socialism and radically democratic governance.[18]

In the last years of the twentieth century, anarchism flowered in ways it had not since 1936. Inspired by struggles such as the Zapatista movement in Mexico, many activists worldwide have become familiar with anarchist tactics and principles.[19] The technology that exposes activists to news and ideas about such movements is anarchistic in nature. This combination of example and techno-ideology fostered a new anarchistic age in which sudden, spontaneous, direct action shut down a meeting of the World Trade Organization in 1999.[20]

Anarchy, State, Utopia

One difference between the common people who transmitted *bruits publics* as tinder for the French and American Revolutions and the drug-addled tricksters who pushed themselves forward as public nuisances in the 1960s was self-identification or ideological articulation. The actors shared an attitude of indifference toward authority and propriety but differed in the consequences of their actions. No self-defined anarchist has ever sparked a revolution. But the ideologically uninitiated who have trafficked in the habits of anarchism—chiefly unmediated communication—have toppled dozens of tyrants.

The historical omnipresence of anarchism, tempered by its general impotence, should have left us with some understanding of its relevance in the world. Yet it remains one of the least understood political and cultural phenomena. Anarchy remains something to fear, something to avoid, something to strive against. It remains a bogeyman, justification for a thousand panics. Journalist Robert Kaplan titled his 2000 book on the future of world politics (with his typical gravity and pessimism) *The Coming Anarchy.* Kaplan predicted that the stable comfort of the modern nation-state is doomed because too many dangerous goods, services, and ideas can flow too easily without the traditional regard for authority and tradition. Especially in areas of the world that lack instruments of civil society, Kaplan warns, "the grid of nation-states is going to be replaced by a jagged-glass pattern of city-states, shanty-states, nebulous and anarchic regionalisms." Drug runners, bands of mercenaries, and weapons dealers will be the real power centers in much of the world, Kaplan fears. This condition, which is more accurately described as "prearchy," "postarchy," or "transarchy," emerges once in a while, most recently in West Africa. It is a transitional state that exists after one mode of authority crumbles and before another rises to take its place. At any local place or specific time, there is an authority. Even a drug-running gang can be an authority. This sort of "government," temporary, arbitrary, and brutal though it is, does not really indicate the presence of anarchy. Yet that is the cartoon version of anarchy that democratic republics have avoided ever since the excesses of French Revolution.[21]

The term "anarchy" has several common uses.[22] To some, anarchy is a state of history or a state of affairs, as in "we are in a state of anarchy." This is Kaplan's point of view. While Kaplan's alarmist descriptions of the collapse of civil society lack sophistication and rigor, he is right to point our attention to the behavior of mobs that destroyed any sense of security or dependability in much of Central and West Africa in the 1990s. These mobs, in the absence of a relevant or dependable state, appear to have governed themselves through communicative channels that were both elaborate and unheirarchical. The raging mob, for nothing yet against everything, represents a sort of "half-baked" anarchism having a negative theory of smashing the state yet lacking a positive theory of how to govern once the state collapses. And it is brutal.

To others, anarchy is a utopian socioeconomic and political vision. Noam Chomsky posits a system of organization called anarcho-syndicalism or sometimes libertarian socialism. Anarcho-syndicalists are more important than ever, not because their vision of society has more traction (although that might be true), but because their chosen tactics have found purchase among a variety of movements and habits and have worked themselves into the toolboxes of many who do not share the anarcho-syndicalist utopian vision. In this third sense, anarchy is a process, a set of behaviors, and a mode of organization and communication. This "information anarchy" is a potentially powerful influence on life in the twenty-first century. Anarchistic habits, structures of thought, matter more to more people every day. Many people around the world have the ability to participate in the "collective anarchistic imagination."[23]

It would be a mistake to overstate the anarchistic effects of global networks. The sophisticated form of anarchy is hardly enabled by the spread of its irresponsible cartoon versions. Global electronic technologies have allowed people seeking to satisfy simple, everyday desires to hook themselves into a dynamic system that dissolves their sense of limits. The act of saying "Why can't we share music with millions of people around the world?" or "Why can't we coordinate mass demonstrations with thousands of people we have never met?" or "Why can't we generate a free and open and customizable collection of software?" has had profound consequences. These shifting expectations have allowed everyday people (albeit technologically proficient and financially privileged) to consider new ways of relating and communicating with one another. Increasingly important areas of life seem outside the reach of state regulation—even if in reality they are not. Revolution does not beckon, but irresponsibility calls and creativity thrives.

CHAPTER 2

The Ideology of
Peer-to-Peer

How did we get from "liberty, equality, fraternity" to "rip, mix, burn?" Sean Fanning, the college student who first cribbed together the pieces of computer code that grew into Napster, is no anarchist. At first he wasn't even a capitalist. Fanning, a music fan and computer maven, was trying to solve two communication problems. The more obvious problem was practical. People wanted to find sites that contained particular songs. Vigilant copyright lawyers had wiped the World Wide Web and file transfer protocol (FTP) sites clean of MP3 copies that were out in the open. Consequently, music fans, hungry to discover music outside their collections, had no online library of MP3s to consult. This was a menial problem, but for digital music enthusiasts, the buzz about the flexibility, portability, and malleability of the MP3 was flowering.[1]

Fanning, perhaps inadvertently, was also trying to solve a much larger communication problem: How do you harness the latent computational and storage power of millions of personal computers into the Internet itself? This is the peer-to-peer challenge: The Internet was designed to be peer-to-peer, but it ceased to be so some time in the mid–1990s. For years the Internet was a conduit for communication among stationary mainframe computers with stable Internet protocol numbers, or addresses. The computers that requested data from the network were the same ones that supplied data. Every Internet user was a both a client and a server. By the mid–1990s, millions of mobile personal computers were hooking themselves up the Internet for fleeting moments. Engineers would have run out of Internet protocol (IP) addresses within months had they not come up with a system that dynamically generates an IP number for a computer on the fringe of the system while that computer is logged in. When the computer severs itself from the network, another may use that IP number. This temporary admission to the network meant that despite growing storage capacity, Internet users generally only took content from the system. There was a high level of interactivity on Web sites, in news groups, and in chat rooms. E-mail (the bulk of Internet use since its inception to this day) operates along client–server principles, even if the data is stored in a remote server belonging to Microsoft or America Online. At least e-mail simulates client–server processes.

Most users are not granted permission to act as servers. Many Internet service providers contractually forbid their customers from using their home computers as servers. So if IP numbers change constantly, and millions of people have billions of files on their hard drives, inaccessible to the curious, isn't there some way to connect these hard drives without going through a gatekeeper like America Online? Isn't there some way of "resolving" a request from an anonymous seeker with a supply from an anonymous donor? Certainly there is. Sean Fanning, like many others, spent some time working on this problem. By the late 1990s, the collective creativity of hundreds of software engineers—Fanning among them—produced some code that solved this communicative problem with an interface that everyday computer users would love. These artists gave intelligence back to the end users of the Internet.[2]

The Nature of Distributed Systems

Most of the public accounts of Napster, Kazaa, Gnutella, and similar digital controversies tend to read like sports articles. Who is winning? Who is losing? They focus on the specifics, the characters, the money, the immediate conflicts. Discussion of the legal cases, when it is good, focuses on the rights and responsibilities of users. But there is far more at stake here than the price of a compact disc, the cost of a lawsuit, or the value of AOL Time Warner stock. Taking the examination to an organic, ecological, and systematic level, I want to start by considering how each case reflects the values and concerns of many well-meaning people—and a few ill-meaning people—on both sides, and how these stories fit into the larger epic of the battle to establish, maintain, rein in, and control distributed information systems, electronic and otherwise. To accomplish this, we must examine the defining characteristics of peer-to-peer communicative systems. Distributed peer-to-peer systems, whether digital or analog, virtual or real, generally share the following qualities:

- They have an "end-to-end" design. All the "thinking" and "memory" of these systems happens at the end point, in many cases a personal computer or perhaps a person.
- They are decentralized. The resources are spread out, making the system invulnerable. An attack on any one part alters the system only slightly. Resources exist throughout the system and flow easily.
- They are antiauthoritarian. There may be guides, mavens, or experts who contribute more to the system than others do. But no one person or committee can turn the system off or remove participants. There is no discernible command-and-control system.
- They are difficult to manage. Removing material or members from a distributed system requires excessive effort and diligence.
- They are "extensible," meaning they support open access to many. People can join with few challenges, and the total number of "nodes" or members may grow exponentially without degrading system operations. "Open access" means it is limited only by ability to work within the protocols. For example, participation in a network of diaspora Chinese communities on the Internet is limited by the ability to read and write Mandarin. Participation in most of the Internet is limited by an ability to speak English and access to an Internet-connected machine. Because of differences in interface design, participation in the noncommercial peer-to-peer system Gnutella requires a bit more technological cultural capital than participation in Napster did.

Distribution (or, to employ an awkward term, distributedness) is not a Boolean quality. It is not either/or, black or white. Most systems that operate in a distributed manner are mixed systems, carrying architecture that has centralized branches in a larger and more open system. For example, proprietary networks like America Online are closed, centralized systems that are linked with the more open and more distributed Internet. The Internet contains many gated communities, and therefore has an increasing number of controls embedded among its architecture. Some systems are more distributed than others. Napster was fairly distributed in that its content was fully distributed among millions of users. But it had a central database of users and a registration process. Gnutella, which has neither, is even more distributed. The Internet was designed to be distributed. But the concerted influence of elites and circumstances have developed a series of tools and strategies that have severely limited its distributedness. Scientific communities are supposed to be open, extensible, and nonhierarchical. But commercial and political interests limit the distributedness of science through formal institutions. Some world religions, such as Islam and Judaism, are paradigms of distributed systems. Others, like the Roman Catholic Church, have highly stratified power and information structures. Many things can be distributed throughout a network or system. But I am concerned chiefly with information in its various forms. Of course, money and music can function as information as well as real, semisolid things.[3]

The Ideology of P2P

Discussions of peer-to-peer systems must get beyond the brand names of networks and the microstory of legal battles and public relations salvos. These stories, though interesting and revealing, are merely bricks in an increasingly important structure—an ideology of file sharing. Let's accept the fact that no company, no court, no technology is likely to shut down widespread peer-to-peer file sharing in the next few years. No new compression format is likely to supplant the remarkably attractive and convenient MP3 format, unless it is just as flexible and easy to use and thus just as threatening to the music industry. No one is likely to regulate bandwidth and access so effectively that Internet users can merely receive and not distribute information—unless most of the electronic media devices in our lives undergo radical changes in structure and function. Although none of these changes are likely or easy, they are all possible. For the sake of exploration, let's assume that the cost and difficulty (politically, economically, and technologically) put them beyond realization in the near future.

What does it mean that more than 77 million people have participated in peer-to-peer sharing? What does it reveal about the ways we share, restrict, buy, sell, and steal information? What are the cultural habits, assumptions, and ramifications of such interaction? In other words, what is the ideology of peer-to-peer file sharing? Many technological theorists, from Martin Heidegger to Jacques Ellul to Marshall McLuhan to Albert Borgmann, have posited that technologies have ideologies or "biases." They may not have used the word "ideology," but that's essentially what they meant.[4] Particular technologies "bias" the judgments or behaviors of those who employ them. This is the matrix of assumptions—the cultural technologies—that real-world technologies reveal, solidify, and extend; they may even determine meaning and intention. Ideologies are structures of language, belief, power, or opportunity that guide thought. They can be actual physical structures. For example, the automobile may have an ideology. Cars embody certain cultural values, such as freedom, autonomy, sex, and rock and roll. They also help create cultural values and phenomena, such as extreme privacy, suburban sprawl, mass consumption, and drive-by shootings. The lightbulb may have one ideology or five. It promotes literacy, domesticity, industry, long work hours, urban safety, and disposability. Combine the lightbulb with the automobile, ramp up their synergistic ideologies to the extreme, and you have Las Vegas.

But what does it mean to "have" an ideology? Does a technology's ideology determine, or at least influence, a culture? Does a culture determine or at least request a particular technology as a way of satisfying preexisting needs, wants, expectations, or demands? Do we want to drive because we have automobiles? Or do we want automobiles because we want to be more mobile?[5] Probably both. Technologies tend to generate or fix values that already flow through some part of a culture, even if they have not reached critical mass. Somebody must have (or predict) a demand for a technology before building and sharing it. Yet few technologies are innocent. Their very presence alters our environment, our worldview. Some become so ubiquitous so fast we can hardly envision a world before them or without them. Try to remember how we maintained social contact before voice mail or answering machines. Better yet, try getting a class full of nineteen-year-olds to imagine an analog world in which they have to flip the record after six songs or walk to the television to change channels, and you'll see what I mean. Yale Law professor Jack Balkin describes ideology as "cultural software." Balkin formulated his theory to explain how ideology guides collective human actions while allowing for difference and awareness of ideological forces. In other words, he was trying to figure out how we can be simultaneously acculturated and differentiated. "Cultural software" is a misleading metaphor, and Balkin cops to its weakness. He doesn't want to offer another cute reductive, mechanistic association between the human mind and computer technology but rather identify what is special about software: its adaptability, customizability, portability, and status as a toolmaking tool.[6]

In Balkin's view, culture is also a technology, or at least it has technologies built into it that guide and influence our social actions and judgments. Like software, culture is malleable and adaptable. So ideology, that collection of toolmaking tools that guide us, can be thought of as "cultural software." Balkin didn't make the leap I am about to make (and might strongly resist it): if technologies represent and extend ideologies, and if ideologies are malleable examples of "cultural software," then some specific technologies (e.g., real software) can be analyzed in terms of their effects on larger ideological systems. I'm interested in how real software influences cultural software. I'm fascinated by how widespread use of distributive communicative technology generates, to employ John Dewey's psychological tenet, "habits of thought." These "habits" among individuals build into "cultural habits," or ideologies, through discussion, deliberation, and distribution.

What about file sharing or the Internet? What is the ideology of peer-to-peer technology? What is the "cultural software" effect of cultural software? Napster revealed the following embedded cultural assumptions:

- Culture is shared.
- Obscurity mimics anonymity.
- Private, individual transactions can't harm large, powerful institutions.
- Local behaviors and actions seem justifiable even at a greater scale and greater distance.
- Large, widespread, uncoordinated actions can't be policed easily, precisely, or moderately.

These cultural assumptions did not emanate from peer-to-peer file-sharing systems. Sean Fanning did not hack this ideology but tapped into values and habits that have existed or centuries. They were not always relevant or important, but articulating them was not hard. Communicative technologies, like many other technologies, reinforce, amplify, revise, and extend their ideologies. By using them, you change your environment. By communicating with others through them, you alter your frames and assumptions about the world. The very presence of a television in a room alters the room's function and structure by changing the perspective of those who inhabit the room. A room without a television, these days, is special because it lacks a window on the Neighborhood of Make Believe. As the San Francisco Diggers of the 1960s used to posit, communicative technologies "create the conditions they describe."[7]

Anarchy of Access Versus the Stability of Ownership

Peer-to-peer technology spreads cultural anarchy when it encourages both "inconspicuous consumption" and "conspicuous production," or, more accurately, conspicuous recombinant reproduction. Peer-to-peer systems, including the Internet itself, are remarkable not because millions of people take millions of files from them but because millions of people compose, copy, place, and distribute millions of files on them. American political economist Thorstein Veblen introduced the notion of conspicuous consumption in his 1899 masterwork, *The Theory of the Leisure Class*. Veblen showed that human beings at the turn of the twentieth century were not rational actors in their consumptive choices but were driven by markers of status that revealed what they valued as a culture and a species.[8] The transactions in peer-to-peer systems, from gossip to Kazaa, are inconspicuously consumptive. No one sees what you get. No one cares who is donating the new files to the system. No one remembers who recombined two or more tracks to make a clever new unauthorized remix. The production and reproduction—the value added—is inconspicuous. Participants are rewarded with cultural capital for donating their (or more often others') contributions to the dynamic communicative process. We are witnessing on a massive scale inconspicuous consumption and conspicuous reproduction. It's the comedy of the commons.

Some have argued (rather naively) that by turning the political economy of communication on its head, such systems offer a fuller realization of knowledge and power for the common good. As philosopher Pierre Levy explains in his book *Collective Intelligence*, the idea captured in his title "is less concerned with the self-control of human communities than with the

fundamental 'letting-go' that is capable of altering our very notion of identity and the mechanisms of domination and conflict, lifting restrictions on
heretofore banned communications, and effecting the mutual liberation of
isolated thoughts." Levy is more optimistic than I am about the emancipa-
tory power of these communicative technologies to bring happiness to the
world. "The problem faced by collective intelligence is that of discovering or
inventing something beyond writing, beyond language, so that the processing of information can be universally distributed and coordinated, no longer
the privilege of separate social organisms but naturally integrated into all
human activities, our common property." We are on the verge of a new age
of communal, global, human consciousness, Levy declares, and we must take
steps to enable this grand new disorder to grow. "To have a chance for a better life," he argues, "we must become collectively intelligent."[9]

This is where the battle lies. The content industries and the states that
support them are trying to perpetuate an ideology as well. Every time film
or music industry people criticize those who share files or those who hack
through technology that would restrict access to cultural expression, I hear
some version of what I call "property talk." This ideology appeals to and
extends the principle that private transactions should be as efficient as possible. The proliferation of efficiency—low costs of transaction and high
prices of remuneration—will generate a cornucopia of choices for consumers. Any function that interferes with private transaction introduces
inefficiencies or leaks into the system. And leaky boats don't float long.
Property talk is "market fundamentalism." For example, Time Warner
CEO Richard Parsons has said, in regard to the proliferation of peer-to-
peer music-sharing networks, "This is a very profound moment historically. This isn't about a bunch of kids stealing music. It's about an assault
on everything that constitutes the cultural expression of our society. If we
fail to protect and preserve our intellectual property system, the culture
will atrophy. And corporations won't be the only ones hurt. Artists will
have no incentive to create. Worst-case scenario: The country will end up
in sort of a cultural Dark Ages."[10]

Property talk is a closed rhetorical system, a specific cultural instrument
that extends a specific agenda or value. Such ideological proclamations accomplish what many closed-system ideologies hope to: They shut down
conversation. You can't argue for theft. This ideology rests on the widely
held assumption that unfettered private control of resources not only pro-

duces the most efficient distribution of these resources but enables some larger public good, such as a proliferation of products and services heretofore unimagined.[11]

We can look at this battle for formats and control as more than a clash over commerce or law. There is an ideological conflict: anarchy versus oligarchy, or access values versus property values. The popularity of peer-to-peer and the resulting public debates have revealed this rift, although it has been largely unarticulated. Many entertain both of these ideologies at various times, even simultaneously. It doesn't take Karl Marx or Jacques Derrida to see that technologies and ideologies create their own internal contradictions, systems of logic that collapse on themselves. In every ideological battle each side needs the other to define itself, even as it defines itself against archaic conditions.[12]

We should strive to pay attention to the ideologies behind the technologies and the technologies behind the ideologies. As world citizens we need to identify the conditions under which the values of property might work best for us and the values of access might best serve our needs. But first we must list and debate our needs. We must then apply each ideology (which is a technology in itself) to these goals and values. And if they come up wanting, we must revise or reject them.

CHAPTER 3

Hacking the Currency

Cybercynicism and Cyberanarchy

Diogenes of Sinope, who lived between 412 or 403 B.C. and 324 or 321 B.C., was infamous. Wearing only a shabby robe, he wandered the streets of Athens and then Sinope after he was exiled for defacing the Athenian currency. He engaged powerful people in playful debate, often exposing their hypocrisy. Asked which city-state he claimed as his own, he responded, "I am a citizen of the Kosmos, or universe." He coined the word *kosmopolites,* from which we derive the notion of cosmopolitanism. Diogenes was a master of satire and parody. He was playful and critical. He loved humanity but refused to grant it the slightest authority over him. Shameless, he knew no manners and lived by his own moral code. He expressed his freedom by masturbating in the marketplace.[1]

If Diogenes wrote anything, it was lost. His legend and teachings lived on through such thinkers as Crates, Seneca, Diogenes Laertius, and Marcus Aurelius. Diogenes was the founder and role model of Greek Cynicism and influenced the Stoics, who provided philosophical guidance to the Roman Republic. The Cynics had no canon, no schools, no academic lineage, but their teachings flowed through the cultures of Greece, Egypt, Asia Minor, and Rome. Those influenced by the Cynics include Christian monastics and Friedrich Nietzsche.[2]

The conventional wisdom about cynicism regards it as corrosive, rude and unworkable, uncomfortable and nihilistic. When I teach about cynicism, I begin by asking my students what they think the word means today. They usually respond that a cynic believes in nothing, is only out for his own benefit, and holds a deeply pessimistic view of human and communal potential. When I ask them to name a popular character who embodies their assumed definition of cynicism, they always cite George Costanza from *Seinfeld*.

The Costanzan version of cynicism is pervasive. But the roots of cynicism are passionately humane and radically engaged. The Diogenic school of cynicism embraces radical individual freedom of expression. It eschews sophistry and theory, and thus presages (and no doubt influences) American pragmatism. Diogenic cynicism values discipline, self-sufficiency, and "living according to nature," or rejecting the influence of social convention or cultural power.

In the film *Casablanca,* when Rick declared, "I'm a citizen of the world—I'm a drunkard," he spoke inaccurately. He was merely a resident of the world, since citizenship demands engagement. By the end of the film, he had become a true and full citizen. Rick achieved state of cynical cosmopolitanism when he embraced a polis of all humanity and reenlisted in the fight against the destructive, greedy, hypernationalistic enemies of humanity. Over the course of the movie, Rick moved from the pretense of a disengaged, selfish manner of cynicism—Costanzan cynicism—to the more responsible and active Diogenic cynicism. Rick and Louis, as they walked over the tarmac of a strangely foggy Casablanca airfield, enlisted in the army of a cynical state. Like Lazlow and Ilsa, they would be warriors for humanity.

Imagine what a cynical state might look like. Imagine what sort of world Diogenes and his confederates would build. Imagine a place where citizens of a borderless polis enjoy total freedom, they organize openly, and communicate freely without regard for structural hierarchy, yet with a deep respect for expertise. Of course, residents of this polis could misbehave, rant, parody, disrupt, and deface currency at will.

What could be a more ideal environment for a cynic than cyberspace—the ethereal realization of cynical politics? Diogenes was a hacker. The Internet is a cynical cosmos. It was designed along cynical principles and serves cynical ends better than any others. And nothing represents the overall nature and substance of the Internet better than masturbating in the marketplace. Where else but the Internet can you book a plane ticket, purchase a book about the history of American copyright, order a bouquet of flowers, write a letter to the editor of the *New York Times,* and wank off—all at the same time?

The word "cynic" comes from the Greek word for doglike. Diogenes, Crates, and other Greek Cynics were accused of living and barking like dogs—shamelessly and uselessly. They, of course, embraced the canine imagery. This gives new meaning to a *New Yorker* cartoon in which a dog, sitting at a computer, says, "On the Internet, nobody knows you are dog."

By taking account of cynicism, by imagining cyberspace as a cynical medium, we can assess the challenges facing us as we try to temper the troublesome behaviors that infect the Internet.

Consider the following questions:

- Why can't we limit access to pornography to adults who view it in private?
- Why can't we prevent massive denial of service attacks?
- Why can't we protect consumer or personal privacy?
- Why do I keep getting these invitations to invest in a business opportunity in Nigeria?
- Why can't anyone make money through this medium?
- Why can't we forge a rich, lively polis out of this powerful communicative technology?
- Why can't we persuade people to respect copyrights in an electronic environment?

Some of the more important questions about the Internet revolve around how people have been using it and to what extent those uses threaten political and commercial interests. First, most American Internet use is peer-to-peer: One person uses it to communicate with another person, largely via e-mail or instant text messaging. Such semiprivate communication is by far the most important use of the Internet. Second, the Internet (through the World Wide Web, following a broadcast model) serves as a source of news, weather, sports, and other information. Web-based commerce is a very distant third and may be slipping. Pornography users rarely cop to their habits when a pollster calls, so there is no accurate way to rank pornography consumption among Internet uses. But the sheer number and persistence of both commercial and free sources of pornography justify citing it among the major uses.[3]

Despite the best efforts of the Clinton administration to make cyberspace safe for commerce, Internet users have not razed the library to build a shopping mall. The power and convenience of e-mail remains the reason new users establish Internet habits. But several important changes in the Internet, chiefly the spread of 128-bit encryption capability in Web site payment systems and Web browsers as well as vigilant enforcement of trademarks, have made commerce at least possible. For most people, the Internet is a combination of three things: a personal communication medium, a library, and a shopping mall. These functions are sometimes incompatible. The full realization of peer-to-peer communication can infringe on commercial efforts to create and enforce artificial scarcity, for instance, when someone copies a news story from a newspaper Web site and sends it to hundreds of others on a listserv or an e-mail list. Posting MP3 files of Beatles songs on a peer-to-peer file-sharing network and thus adding to the library of knowledge available on the Internet also can undermine commercial efforts. Conflicts among functions emerge chiefly when the commerce involves information or culture goods. For products that can be compressed into a tiny digital file, cyberspace offers none of the friction, delay, and inconvenience that the analog world imposes on the information ecosystem. Efforts to regulate the Internet have focused on building in friction, thus privileging the shopping mall at the expense of the communication medium and library functions. Increasingly, the radical freedom of content afforded by the medium and library functions have alarmed guardians of security in the real world and have precipitated calls to monitor and restrict the very activities that drove the spread and adoption of the technology—e-mail and pornography.

Once we recognize that regulating the Internet will be like trying to regulate Diogenes himself, we become much more modest. We must concede that each of these problems can only be approached through deliberation, discussion, debate, and dialogue. Every attempt to guide behavior on the Internet must emerge from a discussion of ethics. Any other methods—technological or legal—may denature the medium and destroy the radical freedom that makes it so attractive and essential.

Since the rise of cynical technology, those who are threatened by cyberspace (or unaware of its nature) have been trying to deny, reconfigure, or rein in its more radical aspects. They are striving to deploy easy technical solutions to a complex medium that—as of today—resists such efforts. Those who wish to quash cynical behavior will have to radically redesign the medium itself.

The Limits of Cynicism

But cyberspace need not embody Diogenic cynicism. If efforts to reconfigure the nature of the Internet succeed, it could embody Costanzan cynicism quite easily. As Lawrence Lessig explained in his seminal work *Code and Other Laws of Cyberspace,* we only assume that the Internet allows radical freedom because it happened to be designed that way. It could just as easily be redesigned to restrict freedom. The story of its design is essential to our understanding of attempts to rein it in.

The Internet was built according to cynical principles—borderlessness, unregulatability, peer-to-peer openness, and peer-review accountability—which also belong to the realms of science and the academy. This is no coincidence. The folks who built the essential protocols of the Internet, from e-mail programs to the server software to the language we use to code Web pages, were by and large academics and hackers, immersed in an ethical environment that rewarded virtue and virtuosity. They were Diogenic cynics.

Arguments that ignore or misunderstand the current "nature" of the Internet abound. In cyberspace, there can be no moderate regulatory moves. Every proposal is extreme. Every attempt threatens to undermine what is good about cyberspace while attempting to squelch the bad stuff. The only way to regulate Diogenes was to expel him.

Misunderstanding Cyberspace

Thomas Friedman, a *New York Times* columnist who specializes in breezily simple assessments of complex topics, wrote in May 2002, "Ever since I learned that Mohamed Atta made reservations for Sept. 11 using his laptop and the American Airlines Web site, and several of his colleagues used Travelocity.com, I've been wondering how the entrepreneurs of Silicon Valley were looking at the 9/11 tragedy—whether it was giving them any pause about the wired world they've been building and the assumptions they are building it upon."[4]

Why didn't Friedman ask executives at Boeing how they felt about terrorists using their inventions to kill 3,000 innocent people? Why didn't Friedman ask Stanley Tools stockholders how they felt about their blades being used by the murderers? The terrorists also used automobiles, photocopy machines, and telephones to execute their plans. Only the cynical technologies attract this level of scrutiny, as if they were the only devices ever put to bad use. In misunderstanding the technology behind the Internet, Friedman assumes that the Web and the habits of the people who use it are special or historically unique. Friedman does not suggest that the terrorists needed the Internet to do what they did. There is much that is new, potentially revolutionary, about cyberspace. But purchasing airline tickets is nothing new. Terrorists have been purchasing airline tickets for as long as there have been airlines. Although the terrorists might have used the Internet and some particular technologies such as encryption and stegonography, there is no evidence that they did. Sadly, the overreactions to the rise of new, strange, and unregulated technologies have distracted us from examining the events of September 11, 2001, with clear eyes.

Using "Our Technology" Against "Us"

In December 2001 in Washington, D.C., the Business Software Alliance sponsored a conference called the Global Technology Summit. The main room for the conference featured a stage not unlike one at Caesar's Palace in Las Vegas. Gaudy columns shoot up into nothing, framing a huge stage backed by multiple video screens. At the podium stood Richard Clarke, then special assistant to the president of the United States for cyberspace security.

Clarke (who resigned from the post in early 2003) is the epitome of the sort of person C. Wright Mills described as a core member of the power elite.[5] A graduate of the Boston Latin School, the University of Pennsylvania, and the Massachusetts Institute of Technology, Clarke started his government service in the 1970s in the office of the secretary of defense. Under four consecutive presidents, Clarke advised leaders on nuclear security and intelligence issues, eventually serving on the National Security Council and becoming President Bill Clinton's chief antiterrorism adviser.[6]

At the global tech summit, Clarke told the audience of commercial software executives, "Our enemies are well prepared to use our technology against us. They understand our technology as well as we do." Since this was the global tech summit, it was not immediately clear what "our" technology was or who "we" and "they" were. But this was a common refrain in the weeks immediately after the terrorist attacks. Such statements drew ethnocentric lines meant to distinguish those who invent things from those who exploit these tools for bad purposes. Such statements ignore the extent to which "we" and "they" are intertwined, interdependent, and in some cases identical. Several of the perpetrators of the attacks of September 11 were educated engineers, trained at institutions in Germany, the United Kingdom, and the United States. If technologies ever have national or cultural pedigrees, they lose them as soon as they are sold or sent to other places and adapted for similar uses. Clarke did not enumerate "our" technologies, presumably tools such as portable telephones, e-mail, the World Wide Web, air mail, fax machines, photocopy machines, credit cards, published airline schedules, box cutters, and jet aircraft. The terrorists used all these things. But Clarke meant to focus attention on the least governable communicative technology available: the Internet.

Clarke identified ten moves the U.S. government and technology companies must make to limit the threat terrorists pose. Partnerships, sponsored research, and widespread public relations efforts should encourage administrators and users to make their networks more secure. Clark endorsed a bill to further weaken the Freedom of Information Act so that companies might discuss vulnerabilities with the government without disclosing proprietary secrets. Much of Clarke's speech rehashed proposals that spread through Washington in the panic that followed September 11, 2001.

His last suggestion really caught my attention. It seemed to be offhand, almost innocuous, but it reveals much about the nature of recent conflicts in the battle to control or unleash flows of information around the globe. "Tenth, we need to look beyond TCP/IP and we need to put money—and again, government and private sector money—into researching the protocols of the future that will deal with the size, speed and the complexity of our future networks."[7]

What is wrong with TCP/IP, the software that makes the Internet work? Clarke did not say. Perhaps he is suggesting that the Internet be reengineered along the lines of a proprietary network. Clarke seems to be complaining that the band of academics and public servants who built the Internet along open protocols can't be trusted to build the sort of system that the U.S.

government and its largest commercial players would like: a more stable, more secure, more monitorable, more controlled network. Not to worry: Microsoft, Cisco, and the U.S. government will build us a better Internet.

Transmission control protocol and Internet protocol are a relatively simple, brilliantly elegant combination of algorithms that fundamentally enable the Internet to be as open and easy to use as it is. They were developed during the late 1970s and early 1980s to manage the traffic of information packets that would be dispersed, sent out along a network, and then reassembled at their destination. TCP/IP manages the dispersal, recordkeeping, and reassembly of packets. The TCP/IP suite has gone through several improvements since the 1980s, mostly through the work of academics and teams of volunteer programmers.

To hear some people talk, the community-based movement produced a static, archaic protocol at the hinges and seams of a rickety global network. On the contrary, there have been constant improvements to TCP/IP. Internet protocol version 4 did the heavy work of building the Internet during its heady early period. Version 6, available since 1995, increases the IP address size to 128 bits, offers better support for traffic types with different quality-of-service objectives, and allows for extensions to support authentication, data integrity, and data confidentiality.[8]

If there were major problems with TCP/IP, if the Internet needed major repairs, energetic thinkers and doers would rush to the job, and this may be what alarms Clarke and the commercial software industry. A sense that the Internet as currently designed is terrifyingly ungovernable has infected the national security culture. Almost a year after Clarke gave his speech in at the global tech summit, White House officials revealed they had considered completely redesigning the Internet and licensing its use.

The Internet is out of control: It always has been. It threatens secrecy. It allows secrecy. It threatens privacy. It enables privacy. It links. It fractures. It allows sexual predators to predate and evade. It lets law enforcement track and trap sexual predators. It conveys offensive material. It allows a free space for people to explore the limits of their tolerance. It forces the use of English. It allows Chinese speakers to share music, stories, and news all over the world. There is a reason for this lack of control. The Internet works so well as an example of a very distributed information system because it was built without "controls" but with "protocols." Let me explain the difference.

Protocols Versus Controls

The first time I heard the word "protocol" in relation to computer technology and communication, I was a newspaper reporter in South Texas covering a hunting trip by President George H.W. Bush. Bush was so un-Texan he shot birds from jeeps to prove his Texanness. I had to figure out how to work an old TRS-80 laptop computer with a phone couplet modem and an eight-line screen. When I finished writing my story, I called the computer technician in Dallas and he talked me through the modem connection procedure. He said the laptop and the mainframe system learned to share data through the use of a "protocol," which he said was like a handshake.

I like that image. A protocol is a handshake. It's a way for different actors to agree on rules of engagement, habits, traditions, or guidelines. If one or the other actor breaks or ignores the protocol, the communicative act fails. If one actor abrogates the terms of a protocol, it will lose the trust of the others. Protocols should be simple enough to allow a diverse array of actors to work with them. They should be flexible enough to allow a variety of interactions over a network or through a system. Controls, on the other hand, are coercive measures imposed by one actor on another. If a protocol is a handshake, a control is a full nelson.[9]

Tim Berners-Lee is the engineer who developed and released such protocols as the universal resource locator (URL), hypertext transfer protocol (http), and hypertext markup language (HTML), all of which enable the World Wide Web to run on the Internet. He did not claim ownership or control of the protocols. In his important book *Weaving the Web*, Berners-Lee explains how he envisioned the Web before it existed. "I told people that the Web was like a market economy. In a market economy, anybody can trade with anybody. . . . What they do need, however, are a few practices everyone has to agree to, such as the currency used for trade, and the rules of fair trading." The protocols that enable the Web simply guide two computers—the client and the server—in how they will take turns sending and receiving and representing information. "When two computers agree they can talk, they then have to find a common way to represent their data so they can share it," Berners-Lee writes.[10]

Most people who enjoy both the freedom and the security of functional democratic republics can agree that both protocols and controls are necessary for society to operate in a way that maintains liberty tempered by predictability, and we should carefully choose when to employ either. The issue is deciding when a protocol is appropriate and when a control is justified. Civilian trials by jury with stable rules of evidence and procedure are protocols. Prisons are controls. A society's charge is to construct an effective and balanced relationship between protocols and controls.

To do this, we need to have an open, balanced, and reasonable conversation about openness and enclosure, protocols and controls, distribution and centralization. But so far, we have only had nasty battles among competing interests. I have taken part in some of these battles with a sense of unease. I hope we can find a better way to order information and global society in the dawning century. The best way to spark such a conversation is to systematically examine the concept of distributed information systems. Distributed systems are all around us and are becoming more important and more prominent than ever.

The most famous battle over distribution involved Napster, the first commercial peer-to-peer file sharing system to achieve global market penetration. It peaked at 77 million users when a federal district court shut it down in the summer of 2001. Napster was a fairly simple software program that enabled an individual to open part of her computer's hard drive to the rest of the Napster community and share digital music files compressed in the popular, convenient MP3 format. Most personal computers can play MP3 files easily, and some inexpensive programs allow users to transfer MP3 files onto compact discs that are playable in standard stereo equipment. The vast library of copyrighted music allowed users to explore music they could not hear on the radio. It fostered better consumer behavior because fans could explore the quality of a variety of songs from an artist before committing money to a compact disc purchase. But the Napster library, like any other library, allowed unethical exploitation as well. Many Napster users assembled complete albums from the individual cuts, and then they burned compact discs instead of buying the commercial disc. There was never a one-to-one correlation between songs downloaded and sales avoided. But the possible erosion of the powerful corporate music industry fostered glee among music fans, anxiety among artists, and panic among music executives, whose revenues did eventually decline due to complex changes in their markets. The record companies sued Napster, and a federal district court issued an injunction against the company for contributory copyright infringement.[11]

As the injunction made it through the appeals process, landing before a panel of the Ninth Circuit Court of Appeals, it became clear that the effects of such widespread sharing on the commercial music market could not be measured or predicted. Some studies suggested Napster retarded compact disc sales. Others argued that Napster sold CDs. The Ninth Circuit could not draw conclusions from these studies and made a rather novel jump. It ruled that because American copyright law allows copyright holders to exploit all possible commercial venues for their products, and because major record companies had been discussing rolling out centralized and controlled digital music distribution systems, Napster had ruined their plans and thus was liable for contributory infringement. The court ruled that a proposed system of control—secure, commercial, digital downloads—trumps an existing system of protocols like Napster, regardless of real harm. In the Napster case, the court allowed one set of controls—the force of law read broadly and absolutely—to enable the implementation of a second set of controls—strong digital rights management schemes that protect digital music files and restrict legitimate use—to stifle competition from a relatively open, decentralized, protocol-based system.[12]

The Napster case was a sign of desperation. If Napster (which had no hope of ever making money despite millions of dollars from investors) prevailed, how could anybody make any money off the Internet? Napster's success would have revealed the fallacies of Web commerce about a year earlier than the stock market did. Purely protocol-based regulatory systems can't enforce artificial scarcity with enough effect to instill confidence among commercial actors. The real question at the heart of the Napster case was: What if no one made money through the Internet?

Debora Spar, a professor at the Harvard Business School, believes that someone has to find a way to make money through the Internet. In her book, *Ruling the Waves,* she explains by historical fiat how this is going to happen. Spar asserts that state regulation of the Internet—the imposition of controls over the protocols—is inevitable as well as justifiable. "Without rules," Spar states, "and particularly without rules of property and exchange, markets simply do not grow." Arguing from a historicist position inspired by Marxist historiography, Spar explains that because earlier unregulated commons like the high seas, the early printing industry, the telegraph, and radio generated anxiety about chaos in their early, heady days, we should consider cyberspace along similar lines.[13] "In the end,"

Spar concludes, "the online music industry may well be regulated by a combination of private and public forces. Firms will set industry standards; governments will enforce them; and even diehard music fans will eventually accept some revised notion of property rights in a digital age."[14]

Spar is half right. Without a strong regulatory system supported by controls and protocols, certain forms of commercial life (not merely the commercial communication that facilitates the delivery of real-world goods) will not take over the Internet. Still, haven't a handful of firms succeeded on the Internet? Amazon.com and ebay are primarily communications media, not electronic businesses. The goods they deliver are analog, solid, and real. The money they accrue is real. The only thing electronic about them is that customers and suppliers communicate with the firms through the Internet. But Spar is mistaken to assume that market forces will necessarily impose order on cyberspace. They could. They certainly are trying. But nothing is inevitable. Just because the high seas were once ruled by pirates and were later tamed by a strong imperialist state does not mean that cyberspace will be as well. History (and states and markets) does not necessarily behave in any one way. The anarchy of cyberspace is no more necessary or inevitable than its potential oligarchy is. As of 2004, both are strong possibilities. But neither is necessary.

What Spar fails to understand (besides the fact that the past does not determine the future and that we impose historical order on events, not the other way around) is that the Internet cannot be regulated in a classic way by simply writing rules and inviting states to enforce them. Its current nature is defined by its openness, its protocol-based structure. Efforts to install friction would change the Internet so radically that it would no longer be the Internet. It might be a useful commercial communicative system. But it wouldn't be the Internet.

Hierarchical controls have been successfully imposed only in the dispensation of domain names. While the regulatory body that governs the domain name system (DNS) has been subject to only symbolic and pesky rebellions, it has been fraught with problems, confrontations, and accusations of antidemocratic and self-serving actions. The Internet Consortium of Assigned Names and Numbers (ICANN) serves at the will of the U.S. Department of Commerce, despite the fact that the Internet owes its allegiance to no single nation-state. ICANN membership supposedly represents all the major stakeholders in the Internet community, including engineers, commercial interests, governments, and a nebulous "public." But ICANN has moved to eliminate its public representation through at-large delegates, and its board has denied a duly elected member access to its financial records. Still, ICANN decides how many top-level domain names (.edu, .com, .org, .uk, etc.) will exist on the Internet and what rules domain-name registration services will follow. It also runs the dispute resolution process that arbitrates conflicts over specific sites (e.g., ruling that rodstewart.com should belong to the rock star Rod Stewart rather than the nonstar Rodney Stewart who registered the domain first). ICANN is dominated by corporate interests (chiefly by favoring wealthy trademark holders over individuals and corporate critics) and stifles democratic accountability. Yet it endures. Although hooligans have occasionally tried to bring down the Internet by issuing distributed denial-of-service attacks on the thirteen root servers that resolve communication (lining up IP numbers like 123.45.67.89 with alphanumeric indicators like http:// sivacracy.net), and competing root server systems have been proposed, ICANN remains the only reasonable solution to Web governance. Its imperfections are apparent and many consider it corrupt. Others consider it impotent. Yet it's all we have.[15]

The late 1990s saw a gold rush for domain names. Squatters and speculators registered names like McDonalds.com in hopes of extorting money from corporations. ICANN was intended to bring order and has. As Harvard Law professor Jonathan Zittrain has declared, however, domain names are not worth much. One may attract readers to a Web magazine called Feed without owning the domain feed.com. Feedmag.com worked just fine. As more people use search engines like Google to find sites, the specific URL matters less. One may issue scathing criticisms of Starbucks or President George Bush without using the domain names starbuckssux.com or georgewbush.org. The text on the page attracts readers. Still, ICANN control over many essential features of the Internet remains the "choke point" of a truly free and unregulated global network. Continuing reform efforts will try to democratize it.[16]

Because regulating a protocol-based system causes problems in the network itself (not least because any revision or replacement of TCP/IP or http demands massive and almost universal recoding), many of the efforts to regulate the Internet have focused on the last mile: the connection between Internet service providers and the end-user computers. The U.S.

Congress has tried to mandate the use of filters in public and school library computers to limit the accessibility of pornographic sites. Because these commercially available filters don't block all filth and inadvertently (or perhaps advertently) block political, health, and human rights sites, they are suspect. Researching the chemical dump cleanup efforts at Love Canal, New York, for instance, is ridiculously difficult in school libraries. In addition, Congress has been considering legislation that would require all digital devices to have copy-control technology embedded in them, thus fundamentally abandoning the concept of the modular general purpose computer. Once again, efforts to narrowly regulate the system to minimize one negative externality would result in a complete and radical rejection of the very openness and usefulness that made the system attractive and important in the first place.

Hacking *The Matrix*

In the spring and summer of 2000, 25 million people downloaded Napster software and used it to share thousands of compressed music files. In the spring of 2000 the film *The Matrix* was released on digital video disc. These two events are not unrelated. *The Matrix* was a major hit movie in the summer of 1999. The film presents a dystopian vision of a technocratic dictatorship that uses information as a weapon of enforcement. Its enforcers are robots who resemble FBI agents. Its rebels are a small confederation of hackers who know the truth about an information ecosystem that has blinded millions of dormant human beings. In the pre-*Matrix* world of the early twenty-first century, our rebels are also hackers. A nineteen-year-old hacker designed Napster as a way to share music without asking permission from music companies or paying them a fee.

The summer of 2000 saw a federal trial in which motion picture companies—including the one that released *The Matrix*—got a preliminary injunction against a hacker journal that shared information on a method to circumvent the access control and anticopying feature of digital video discs. Hackers wanted to see movies on their Linux-based computers. To see *The Matrix,* they had to hack the matrix.

The Matrix quickly entered the public consciousness. Teens used its phrases in conversation. Fashion trends changed: wraparound sunglasses, long leather coats, and chrome mobile phones proliferated. The film's amazing special effects and elaborate mythology entered the public imagination. The political messages of *The Matrix* were hard to miss: unbridled technology is seductive and potentially destructive, information can be a weapon of oppression as well as liberation, and hackers are heroes. The hacker ethic rests on openness, peer review, individual autonomy, and communal responsibility. Anarchism built the Internet. But the threat of anarchy has launched a decade-long effort to rule it and rein it in. The outcome of this battle is far from clear, but the battle itself has damaged the progressive potential of this powerful communicative network of networks. Every time we rein in the lesser form of cynicism, we irreversibly damage the greater, more valuable form as well. Diogenes still thrives on the Internet, but Constanza is alive and well in the real world.

CHAPTER 4

The Peer-to-Peer Revolution and the Future of Music

On a Wednesday afternoon in 2002, hours before the president of the Recording Academy used the 2002 Grammy Awards ceremony to berate music fans for downloading billions of music files from various peer-to-peer systems, I bought four compact discs and two boxed set collections—nearly $120 worth of music.

Why do I still buy recorded music? I ask myself this question every time the credit card bill arrives. I also ask it every time someone from the big record companies complains about decreasing music sales. I personally spend hundreds of dollars per year on compact discs and legitimate, commercial digital downloads. I should be the record industry's dream customer. Instead I'm a thief: I download thousands of songs per year from Gnutella, a noncommercial peer-to-peer file-sharing system built by volunteers who love music as much as I do.

On that Wednesday, in the middle of the awards show, Recording Academy president Michael Greene blamed the industry's woes on folks like me. He said it had been a tough year for music sales because of "the unbridled advance of the Internet." Among all the problems facing the industry, Greene said, "the most insidious virus is piracy on the net." Greene claimed that "ripping (converting digital files from compact discs into portable MP3 files on a computer) is stealing [artists'] livelihood one digital file at a time, leaving their musical dreams haplessly snared in this World Wide Web of theft and indifference."

Greene's tirade was not out of step with messages the industry as a whole was sending. His complaints, rude as they were, raised important questions. Does each downloaded song equal a lost sale? I am not convinced.

I would not consider buying ninety-nine out of one hundred songs I download. Many of them are garbled, low-quality recordings. Some are partial files. Some turn out to be lame songs. Mostly, I download songs to see if I want to buy them. I also get songs I can't buy, for example, out-of-print stuff from cult bands like Too Much Joy or rare live cuts of "Jersey Girl" sung by Bruce Springsteen and Tom Waits. Clearly the music industry has lost sales in the years since it killed off Napster. Is this lull a historical anomaly? Is it the result of alienating consumers? Is it the result of fewer titles for sale? Are compact discs overpriced? Do consumers have fewer marginal entertainment dollars to send to the recording companies, and do they have more entertainment options? If downloading is pervasive and detrimental to music sales, why aren't these companies doing worse than they are? Why does anyone who has a networked computer buy music at all?

Every time I think about Greene's speech, I ask myself, Am I a thief? I justify my downloading habit on the ground that I buy what I like and throw back the small fish. So in the long run, I do no harm. Back in the analog world, I did this sort of thing for years. I came of age with vinyl LP records and flimsy cassette tapes. I recorded *KISS Alive II* from my friend Larry's record collection back in 1976. Other friends gave me tapes of the Clash and Elvis Costello, and I gave them tapes of the Jam and Joe Jackson. Later I replaced my tape collection with vinyl albums. If I could find a good used copy, I bought it. If not, I bought a new one. Like most students, I lacked disposable cash, but what I had went for music. So was I a thief all along? If I cared less about music, I would have recorded fewer cassettes, but I would also have purchased fewer albums. Before the rise of peer-to-peer music distribution, I don't remember anyone asking these questions. I certainly never asked them about myself. I lived with music however I could get it. Sometimes I paid, sometimes I didn't. No one seemed to care one way or the other.

Clearly the anarchistic digital world has altered our ethical scaffolding and cultural habits. Even more significantly, there seems to be a serious abrogation of the presumed contract between a significant percentage of music fans and the companies that distribute music. Which side broke the contract? When did we sign it? Is copyright law at its foundation? Perhaps the contract itself is obsolete, or at least misplaced within cultural economies that involve such slippery, malleable, portable products.

This consumer behavior is merely an amplification of ancient habits that predated the brief rise of a commercial music industry in the twentieth century. We share music in a circle. Music doesn't work the same way for us if its piped to us, processed for us, and stored at a clean distance. We want to groove to it, mess with it, remake it. We want to make it ours and use what flows around us to build new music. And, as always, we want to pay as little as possible to be able to play and play with music.

Unfortunately the urge to share does not please firms that produce and distribute the world's best-known music. The rise of peer-to-peer technology highlighted these cultural habits in a way that threatened the powerful companies that invest billions in production, distribution, and marketing. They are fighting back with lawsuits, propaganda, and restrictive technology. While many users insist that music is special to the human community and thus deserves special rules and considerations, those who invest time and money in the industry insist that a product is a product. In their view music deserves a high level of protection against unauthorized exploitation. The big question is, Can we enjoy the benefits of large, successful, commercial music companies and easy-to-find free digital music files? Must we choose? And if so, who gets to do the choosing?

A Sports Story and a Crime Story

At the Napster trial in 2001 both sides introduced shoddy studies purporting to show the effect that 70 million people sharing songs would have on the music market. Neither side convinced the appeals court of its position. Perhaps Napster retarded sales. Perhaps Napster sold discs. Perhaps it decreased sales of some artists and increased sales of others. We still don't know. But examining the political, technological, economic, and cultural dimensions of the battle over music can reveal some interesting things about the ways we use music and what we expect from our media and cultural systems. By considering the cultural effects of free music flowing around the globe, we may learn something about the stability of cultures and the economic constraints that have rendered them mutually inaccessible.

Here are the basic facts of the crime story in progress. The commercial music industry took in more than $33 billion worldwide in 2002. In calendar year 2000, the only complete year in which Napster operated, compact disc revenue was up 3.1 percent from 1999. In 1999, the year Napster debuted and MP3s became widely available through various other means around the Internet, compact disc revenues were up more than 12 percent. In 2001, when the music industry successfully shut down Napster, revenues were down 4.1 percent. In 2002 sales of compact discs fell by 9–10 percent. In 2003 the drop mimicked the slope of 2002. Still, sales and revenues in 2003 were comparable to 1997, suggesting that the massive sales increase in 1998, 1999, and 2000 may be the anomaly. Meanwhile, peer-to-peer usage surged through these years, with some reports claiming that more than 230 million people worldwide had installed some form of peer-to-peer client software by the summer of 2003. Kazaa software may be the most installed network application in the history of personal computing.[1]

Most of what was written about Napster in 2000 and 2001 read like a sports story: Who will win? Who will lose? What will happen to poor Metallica? But soon after the initial shock and buzz, music fans and technology experts began sensing that the legal and financial bickering obscured something more important: imminent changes in how people share, restrict, and distribute information and cultural products in the twenty-first century—the potential "Napsterization" of just about everything. Yet Napster's rate of adoption was astounding. By the time Napster went dark after a court order in the summer of 2001, more than 77 million people had registered with it—more than twice the number who used America Online in 2001. Yet the company never charged a dime for its service and never generated a way to make any money from anyone. It burned through more than $25 million in foolishly granted venture capital funding. Once it ran out of money and legal arguments, all it owned was its trademark and a database of former users.[2]

After Napster went quiet, several other peer-to-peer services sprang up. Some, like Kazaa, are run by companies and are likely to be shut down. Others, like Gnutella and FreeNet, are nonproprietary, fully distributed, and seemingly unstoppable. Millions of people trade billions of songs every month on these services. Their popularity has surpassed Napster's. Had the record industry cut a deal with Napster, it might have avoided the ungovernable chaos of decentralized peer-to-peer services now taking over the Internet. But it did not. Because the demise of Napster did not stem the tide of file sharing, the music industry shifted its tactics and targets from peer-to-peer services to Internet service providers, colleges and universities, and the compact disc itself. Meanwhile, the industry tried to interest consumers in pay-per-song services. The recording industry's reaction to the rise of peer-to-peer services demonstrates a fervent disrespect for music fans. Through 2002 and 2003 it hinted at "making examples" out of a few, yet it has consistently berated all. The industry has been trying to take the anarchy out of the music, and take the music out of the anarchy. Failing on all other fronts, the music industry in the summer of 2003 started filing civil lawsuits against several hundred individuals. These suits embarrassed the industry and alienated the public. By 2004 there was little reason to believe that the lawsuits would stem the demand for open libraries of music.

I suppose the recording industry thinks it can get away with such tactics because it used to exercise almost complete control over how we use music in our lives. Peer-to-peer threatens its monopoly on marketing and distribution, if not revenue. It introduces uncertainty and unpredictability into an already unstable market. To reach fans, artists no longer have to go through the labels. To hear cool music, fans no longer have to turn on MTV or listen to a narrowing array of radio programs. There is no telling what fans will find, and perhaps no clear feedback to the labels about what people want. Importantly, the industries invested in the distribution of music are broad and include more than the major commercial recording companies. Although music fans are spending less on compact discs, our music habits now enrich companies far beyond the major labels. Every "thief" who downloads MP3s and buys compact discs is also a dream customer for makers of compact disc burners, portable MP3 players, broadband Internet access, and faster personal computers. Nothing is free. The question is, Who will get paid?[3]

Anarchy in the PC

The following advertisement, titled "A Big Fat Thanks to Record Execs," ran in the *New York Times* business section in October 2002. It was a simple, scribbled text, presented as if written with a Sharpie marker:

Thank you for fighting the good fight against Internet MP3 file-swapping. Because of you, millions of kids will stop wasting time listening to new music and seeking out new bands. No more spreading the word to complete strangers about your artists. No more harmful exposure to thousands of bands via Internet radio either. With any luck they won't talk about music at all. You probably knew you'd make millions by embracing the technology. After all, the kids swapping were like ten times more likely to buy CDs, making your cause all the more admirable. It must have cost a bundle in future revenue, but don't worry—computers are just a fad anyway, and the Internet is just plain stupid.

The ad was placed by *Rolling Stone,* the once revolutionary (now generally stale) music fan magazine, and revealed widespread attitudes about peer-to-peer file sharing over the Internet. The ad makes two assumption: (1) the major labels are not serving the most active and passionate segments of the American (if not world) fan communities and (2) music fans who download songs "spread the word" about new music, and without peer-to-peer accessibility, such conversations would stop. Neither assumption is necessarily true or false. Before digital peer-to-peer networks, music communities spread the word about new, exciting artists from peer to peer, like the Parisian commoners gossiping in Luxembourg Gardens. This anarchistic fan culture was essential to the germination, development, and proliferation of genres that were once countercultures but now make up significant segments of the commercial music catalogs such as hip-hop and punk rock. Both of these genres, which emerged from 1970s cassette culture, were anarchistic through and through. The movements had no leaders, although mavens soon rose from the ranks. They started and thrived among economically and socially disaffected urban youth in the United Kingdom and the United States. They forged infectious works that soon crossed the globe via unanticipated routes, following the flows of tourists, fans, and mail. Neither movement had any interest in the property or propriety standards of the majority culture and its corporate institutions. Both were politically engaged in the beginning. Youth culture is anarchy. Punk, in fact, was overtly and unashamedly anarchistic. The punk movement's banner song was the Sex Pistols' "Anarchy in the UK."[4]

The *Rolling Stone* advertisement sat just beneath the *New York Times'* listings of the top-selling media products in the United States. The top-grossing film that week was *Jackass: The Movie*. The top ten music albums included such established acts as the Rolling Stones, Elvis Presley (both for greatest hits collections), Faith Hill, and LL Cool J, as well as artists who appeal almost exclusively to young consumers and emerge from (while not necessarily carrying on the deep traditions of) anarchistic youth culture genres: Avril Lavigne, Nelly, and Eminem.[5]

The success of Eminem's hit album of 2002, *The Eminem Show*, is instructive. Interscope Geffen A&M, Eminem's label, released his album several weeks before its planned June release date because someone had leaked tracks from the album over peer-to-peer file sharing systems more than a month early. The label had gone to great lengths to prevent reviewers from ripping review copies of the compact disc into downloadable files. Instead of mailing out discs several months early, the standard practice for decades, the label threw listening parties for journalists at the offices of Vivendi Universal, the conglomerate that owns Interscope Geffen A&M. In addition, the label released phony tracks over peer-to-peer networks that contained only snippets of the songs looped repeatedly.[6]

Yet someone got hold of the entire disc and released the files to the public. A few days before its official release *The Eminem Show* reached the number two position on a list compiled by Gracenote, an on-line service that searches for artist and track names when people play compact discs in computers that are connected to the Internet. The ranking indicated that thousands of people were playing compact discs that had been burned with Eminem's unreleased music.[7]

Given the widespread availability of unauthorized digital tracks in MP3 form and privately created compact discs with many if not all of the songs from *The Eminem Show*, the market for the disc could be expected to suffer. By the summer of 2002, about 100 million computer users had file-sharing software on their computers. But there is no evidence that prerelease distribution did much to retard sales. In the first three weeks consumers purchased 2.4 million copies of the disc from retailers. More than 1.32 million albums moved in the first week alone. In contrast, Eminem's second album, *The Marshall Mathers LP*, sold a healthy 1.8 million copies in its first week during 2000 and ultimately more than 8.7 million discs, back when about 77 million people were using peer-to-peer music-sharing networks. Limp Bizkit, Britney Spears, and *NSYNC, acts that chiefly appeal to suburban young Americans, have joined Eminem as the only pop music acts ever to sell more than 1 million albums in their first week after release. And they all accomplished this feat during an era of rampant file sharing. *The Eminem Show* was the top-selling album of 2002, with consumers buying 7.4 million copies in the second half of the year alone. Eminem's success in 2002 stands out in an otherwise dismal sales year for the music industry. But if downloads were to blame for the sales slump overall, why wouldn't the most downloaded album suffer? How could *The Eminem Show* dominate the charts after its songs were distributed for free?[8]

A Dying Industry?

Major recording labels perform four basic tasks: production, distribution, price fixing, and gatekeeping. But bands today can use home computers to record, mix, and edit their music, put up their own Web sites, or give away their MP3s on peer-to-peer systems. They make money through live appearances and the sale of T-shirts and other paraphernalia. So production and distribution don't seem so hard anymore. The new technology evades the professional gatekeepers, flattening the production and distribution pyramid. It's risky. It's not lucrative. But being a musician has always been risky and hard. The artificial space between creators and consumers is closing in terms of the feedback loop and the professional identity of creators. Digitization and networking have democratized the production of music and allow more immediate feedback between artists and fans. Artists are more vocal than ever about the ways the commercial music industry has exploited their work. So the commercial record industry is experiencing pressure from its consumers and its labor force. Neither supports the way the business has been run, and both are threatening to take more control of the system. The anarchists are at the gates.[9]

Throughout 2002, the music industry complained about slipping compact disc sales, down by 8 to 11 percent. By the end of the year most reports arrived at a consensus that the major labels sold 9–10 percent fewer compact disc albums in 2002 than in 2001. Similar figures ran through 2003. The music industry and thus the reporters who generate and perpetuate conventional wisdom about the future of the industry declared that downloading caused this loss. But the story isn't so simple. Careful examination of the record does not reveal a simple relationship between song-by-song downloading and album sales on a compact disc. Although many experts and interested parties in the music industry declare the widespread habit of sharing music an emergency, some have provocatively argued that such sharing actually generates legitimate music sales. A study released by the market research firm Forrester Research, Inc., concluded that 31 percent of music consumers download music and burn CDs often. These same digital music users buy 36 percent of all CDs. Downloaders are the music industry's best customers. The Forrester report blames extra-network factors for the drop in compact disc sales, including a worldwide recession in the early 2000s, fewer titles released, and a rollback of teen superstar acts that had caused a spike in the years immediately preceding the survey. Perhaps one of the most revealing statistics about the sales of compact discs came from the recording industry itself: The industry released around 27,000 titles in 2001, down from a peak of 38,900 in 1999. Since year-on-year unit sales dropped a mere 10.3 per cent while the supply of new titles was down about 25 percent, average sales per title have gone up. It is reasonable to conclude that there were more factors at work than disc sales being replaced with song-by-song downloads.[10]

Stanley Liebowitz, an economist at the University of Texas–Dallas, asserts that the future of the music industry is not clear. His conclusions are frank and open to revision, and sometimes Liebowitz seems to be the only open-minded person speaking about the issue. He studied the reports of sales slumps that the music industry was circulating in the middle of 2002 and concluded that the claims were overstated. If we discount the sales lost from discontinued cassette singles and compact disc singles—moves long planned—sales were down by only 6 percent between 2000 and 2001. Such a drop was not unusual for a diverse industry in a recession. About halfway through 2002, Liebowitz looked at the historical data and concluded that the slump in sales coincided with previous macroeconomic downturns. Six months later he saw the data for the last half of 2002 and concluded that these were indeed extraordinary times for the recording industry. Most significantly, he discounted any claim that the price of compact discs had risen above the rate of inflation since their introduction in the mid-1980s. The proliferation of peer-to-peer usage and the availability of compact disc burners might be a major (if not the major) cause of this downturn in sales in the recording industry.

Certainly consumer alienation plays a part, as do consumer spending habits in general. Something is happening here. But we don't know what it is. There is a very good chance that these conditions are probably contingent on a wide array of independent variables. The industry could change its methods. Consumers could change their habits. There are very good reasons to believe that we will not have to choose between life with peer-to-peer or life with major recording labels.[11]

The Perfect Jukebox

The Recording Industry Association of America (RIAA) certainly believes that the public is facing a Boolean choice: one system of distribution or the other. But why is the music industry so alarmed? Why can't it recognize that sales drop off and music leaks through, as they always have, only more so? Why can't it offer a diverse array of services over and above the stream of digits, the notes and chords? What is the source of the moral panic that has overwhelmed discussions of how we share and enjoy music in the twenty-first century? In fact, the rise of peer-to-peer systems and MP3 compression destroyed a dream, not a reality.

The dream was to build a global jukebox. It was going to offer us almost every song at almost any time. Once the infrastructure was built (and much of it already existed), distributing each song would cost the industry almost nothing. No one would be able to copy these songs without going to extreme measures. This dream, of course, was not exclusive to the music industry. One of the central principles of information policy during the 1990s focused on the idea of bringing smaller and smaller "works"—a song instead of an album, an article instead of a magazine, a single film in lieu of a movie channel subscription—to consumers. In turn, consumers would pay a flexible, metered price for access to the materials. For those who watch only two films per month on their televisions, paying $5 per film seemed a better deal than paying $15 per month for a subscription to a movie channel. Likewise, paying for a daily assemblage of news from a variety of sources focusing only on subjects that would interest a consumer seemed a more efficient model. Consumers would not pay for unwatched or unread content. Producers would have better market signals that would guide their production choices. More importantly, distributing arrays of data is marginally cheaper than compiling and producing a larger physical work—at least that was the assumption in those heady years.[12]

The only thing standing in the way of a controlled, protected, encrypted system of commercial digital music distribution was agreement on standards and methods. The music companies had to come up with a way to implement some form of digital rights management (DRM). The producers of music and the hardware that plays music had to concur about methods and standards so that consumers would not have to play Sony discs on Sony players only. All authorized commercial files had to play on all authorized commercial systems. And no unauthorized files should be playable on any authorized hardware. To accomplish this, the dreamers would have to radically rewrite copyright laws.

They did this in 1997 and 1998 by effectively outlawing software that could evade or break through digital rights management schemes. Then, because digital networks were global while legal regimes were national, they had to spread such standards across the globe through international rule-making bodies. The globalization of content control continued through 2003. And closer to home, they had to get all the content producers in a given industry—software, music, video, text—to concur on technological protection standards so that media products from multiple sources would work on most of the machines available in the market. And they had to get the hardware manufacturers to agree to those standards as well. The campaign stalled.

That's the vision the Secure Digital Music Initiative (SDMI) was supposed to fulfill. Flush with unwarranted confidence from the passage of the Digital Millennium Copyright Act, the big music companies got together to figure out what the encryption standards would be, which digital players would be allowed to license the standards, and the methods of warehousing and distribution. The result of the SDMI would be a large array of licensed single-song files available for download from a handful of sources on the Internet. These files would not be transferable from the computer to which they were downloaded, would not be usable on portable digital music devices, and would not be burnable to compact discs. Not surprisingly, no consumers were invited to the SDMI deliberations. The wants and needs of consumers seemed antithetical to the effort. Disrespect for music fans seemed to be one of the initiative's prime motivations.[13]

The RIAA announced the formation of the SDMI in late 1998, but the dream had formed in the minds of record industry insiders much earlier. About 180 companies from all corners of the globe eventually contributed to the consortium, representing major segments of the music, electronics, and software industries. Without a specific timetable or agenda, the SDMI set about to make the Internet safe for controlled music in two stages. First it would agree on some watermarking process and require its manufacturers to produce digital music players that would play these files easily, so that the digital rights management scheme was thin enough to seem invisible to the user. Second, the SDMI would work on standards to make these new players—whether hardware or software—reject any music file that arrived without the requisite watermark. The plan was to make cool music so cheaply and easily available that consumers would fill their lives with the new devices and grow comfortable in the new, highly regulated music environment.[14]

But they never really got to the first step. There were three problems with this scheme. First, the representatives and engineers at SDMI could not agree on much. As journalist John Alderman relates in his book *Sonic Boom,* SDMI was shaken by discord and miscommunication between record industry people and engineers. Company representatives to SDMI could expect many pleasant junkets to sites like Tokyo, New York, and Florence, and they accomplished very little very slowly. The SDMI effort stalled in 2001, as it became clear that consumers were very much in love with the relatively open MP3 format for music files and had the tools to convert CDs to MP3s. No one came up with a protection technology that was hard to crack. No one could get hardware manufacturers to get on board. Still, the effort to protect music with digital rights management schemes lurches on—sometimes tragically, sometimes comically.[15]

The music industry has been unsuccessfully trying to stall the expansion of unregulated distribution of content through litigation until it can establish a standard secure digital encryption format, which is an essential step toward a global "pay-per-view" system—a proprietary information ecosystem. Consumer activists and librarians have been warning for some time that such technocratic regimes could be a severe threat to democracy and creativity around the world. They involve the efforts of the content industries to create a "leak-proof" sales and delivery system, so they can offer all their products as streams of data triple-sealed by copyright, contract, and digital locks. Then they can control access, use, and ultimately the flow of ideas and expression. The content industries have been clear about their intentions to charge for every bit of data, stamp out the used CD market, and crush libraries by extinguishing fair use. In early July 2000 America Online signed a deal with a digital rights management system called InterTrust to provide encryption and decryption technology to AOL software so that AOL users would endure metered and regulated use of digital music film, text, and everything else. This has yet to happen. There is a strong sense that consumers would not stand for it. Other digital music services have been struggling to negotiate licensing terms with the record industry so they can partner to install electronic controls on their music.[16]

The culture industries can take advantage of the "digital moment" to trump the democratic process and write their own laws because digital formats collapse the distinction between using and copying material. The regulation of reading or listening raises deep First Amendment concerns, and courts have been unenthusiastic about limiting these activities. However, copyright law regulates copying, and so digital distribution allows a higher level of regulation than anyone ever imagined. If the day ever comes when we have to apply for a license to listen or read, content producers will be cops, juries, and judges in matters of copyright.[17]

Music for the People or by the People?

Despite efforts to dictate format, the people have spoken. They want their MP3s. MP3 is a "codec," or coding-decoding algorithm, that compresses complex music files into relatively tiny files by removing signals the human ear can't sense under most circumstances. One MP3 file is still big enough to make distribution over phone lines and 56k modems excruciatingly slow. But increasing numbers have access to high-speed Ethernet, digital subscriber lines (DSL), or cable modem connections at home, work, or school. Such large numbers of MP3s travel through these high-speed networks that they sometimes slow to the speed of a 56k modem connection, which generates massive headaches for system administrators in companies and universities.[18]

Designed with consumers in mind, the MP3 format is remarkably easy to use, share, copy, edit, filter, and mix. People can convert old vinyl collections to newly portable MP3s. They can mix tracks and beats. They can clean up scratches and pops in old songs. They can burn new compilation CDs, just as they did with cassette tapes for nearly thirty years. And MP3s sound almost as good as CD files, especially through small computer speakers. Only the most discriminating audiophile can tell the difference. The MP3 format may not have satisfied aficionados of the Kronos Quartet, but it was good enough for punk rock. And those infused with the spirit of punk rock—a larger number every day—embraced the format. It worked.

What destroyed the dream of SDMI was Napster, which turned the dream of a global jukebox into an anarchistic nightmare. Consumers and music fans figured out how to hot-wire the jukebox. They would get their music song by song—on their own terms. In desperation, the SDMI decided in late 2000 to issue "an open challenge to the digital community." The industry invited computer security experts to crack the copy protection technology SDMI was considering. It would be like a field test of defenses. In the first round, the SDMI offered four watermarking schemes and two technologies that would govern how users could reuse and remix tracks on a compact disc. A watermark is a small piece of code that certifies the authenticity of a digital file. It might be impossible to play a file without an authorizing watermark on a watermark-sensitive machine. In this case, a file with a watermark would warn a copying machine such as a computer or MP3 player that the file should not be played except under specified conditions. The SDMI offered a cash prize for successful cracks if the participants agreed not to disclose the methods they used to defeat the system (generally the DMCA prohibits this sort of research, but the contest specifically authorized the efforts as an exception to the DMCA). A team of encryption experts led by Edward Felten at Princeton quickly defeated the four watermarking schemes. They removed the watermarks without degrading the sound quality of the music files. Because the team was not interested in the cash reward, it did not agree not to disclose the information. In fact, the academics on the team recognized an ethical duty to disclose the results of their study.[19]

As the research team was preparing to release its findings at an academic conference in the spring of 2001, the RIAA sent a threatening letter to Felten. "On behalf of the SDMI foundation," the industry lawyer wrote, "I urge you to reconsider your intentions and to refrain from any public disclosure of confidential information derived from the Challenge and instead engage SDMI in a constructive dialogue on how the academic aspects of your research can be shared without jeopardizing the commercial interests of the owners of the various technologies. . . . Because the public disclosure of your research would be outside the limited authorization of the agreement, you could be subject to enforcement actions under federal law, including the DMCA." Felten decided not to subject his team, which included graduate students, to a long legal battle over the reach of the DMCA and withdrew his paper from the conference. Sometime later, after the RIAA and SDMI succumbed to public pressure and ridicule, Felten and his team delivered the results of the study. This story had two morals, both of which guided how efforts to market and control music would play out over the next two years. First, digital rights management schemes are easy enough to break that they don't deter the serious or passionate; and second, computer scientists and academics would face intense pressure from powerful interests if they followed the ethical directives of their calling and subjected their own work to open peer review. The chilling effect worked both ways. But one side had more expensive lawyers.[20]

The Battle Moves from the Courts to the Computers

Holger Turk is a music fan in Germany. In 2002 he bought a copy-protected compact disc called *Through the Looking Glass* by the group Toto. On the label were printed the following words: "It is designed to be compatible with CD audioplayers, DVD Players, and PC-OS, MS Windows 95, Pentium II 233 Mhz 64 MB RAM or higher." When Turk tried to play the disc, he found those claims to be false. The disc did not work on his DVD player or his computer. So he wrote to EMI, the company that released the disc, to complain the disc was worthless. A public relations representative of the company wrote back:

Dear Mr. Turk,

We will spare ourselves the trouble of addressing those observations in your email which are obviously uninformed. Simply realize: more than 250 million blank, recordable discs and tapes were sold and used this year, in comparison to 213 million prerecorded albums. . . . The widespread copying of prerecorded audio material via the burning of CD-Rs can only be countered one way: namely, copy protection. . . . In the event that you plan to protest future releases of copy-protected CDs, we can assure you that it is only a matter of months until more or less every CD released worldwide will include copy protection. To that end, we will do everything in our power, whether you like it or not.

Sincerely,
Your EMI Team[21]

One of the greatest successes the music business had in the twentieth century came back to haunt it at the end of the century. In the 1980s, record companies persuaded millions of fans to repurchase their music collections on the new digital compact disc format. Back catalogs, long thought exhausted of commercial potential, boomed again. Everyone needed a new copy of the Beatles' *Rubber Soul* because their old copy had strange splotches on it and "Norwegian Wood" had a huge scratch. Compact discs, consumers were told, sounded cleaner and brighter, and lasted longer. They were light enough to run with, convenient enough to use in a car, stable enough to withstand dozens of people dancing on an old hardwood floor. The only things consumers had to give up were the ritual of interrupting a kiss to flip to side two of *Led Zeppelin II,* the working zipper on the cover of the Rolling Stones' *Sticky Fingers* album, some warmth and character from familiar tunes, and from $12 to $18 per compact disc. The compact disc seemed like a really good idea at the time. But the music industry was offering its entire catalog in unencrypted digital form. Short of selling pirate copies of these discs, there was no contract, computer code, or law that could prevent consumers from doing whatever they wanted with their new digital libraries. As soon as consumers bought good enough speakers for their increasingly powerful office and home computers, they found it useful to store their entire digital libraries on their ample hard drives. Soon they wanted to share.[22]

After regretting this strategic mistake, abandoning SDMI, and vanquishing Napster in court, the recording industry embarked on a campaign to recover control over the distribution of music. First, it introduced encrypted and enhanced compact discs that offer extra features like video clips and links to Web sites that offer discount promotions. Most of these protected discs were marketed in Europe. Only a handful made it to American stores. Of course, technical failure and the resulting consumer frustration—not to mention the nasty attitude of record companies like EMI—threatened to alienate consumers from the disc format altogether. So by early 2004, the campaign to build a better compact disc was stalled.[23]

In 2003 the RIAA sent letters to colleges and universities, warning them that they have a legal and moral responsibility to ensure that no one is using their networks to distribute copyrighted material without authorization. Meanwhile, two companies that the recording industry had hired to scour the Internet for violators were barraging university technical support staff with "cease and desist" letters. A group of university presidents reacted to RIAA pressure by complicating the issue. The presidents said they would "seek ways to reduce the inappropriate use of P2P technology without restricting free speech and expression, invading privacy, or limiting the legitimate uses" of the technology. It was not immediately clear how an institution might install controls that would stem the sharing of data without massive reengineering of the communicative systems themselves. The universities are charged with balancing interests and values. So they did what universities do best: They thought about it. While traditional academic institutions pondered, the U.S. Naval Academy acted on a tip. Empowered by the "honor code" that all midshipmen must sign, the academy seized one hundred computers from students suspected of downloading copyrighted material. While the midshipmen faced the military version of due process, they lost access to their computers, hard drives, and personal archives for some time.[24]

Undaunted by the university presidents' concern for academic freedom and privacy issues, the RIAA launched a full assault on the privacy of high-speed Internet users by increasing the pressure on Internet service providers (ISPs). In January 2003, the RIAA persuaded a federal judge to force the telephone company Verizon to reveal the name of one of its high-speed digital subscriber line (DSL) customers who was offering and retrieving songs at what the RIAA claimed was a rate of six hundred files per day. Verizon immediately appealed the order to a higher court, arguing that it has a contract with its subscribers to protect their privacy. The RIAA had not filed a lawsuit against the DSL user. It only wanted her or his name so it could consider pursuing legal action later. Under the 1998 Digital

Millennium Copyright Act, ISPs must remove suspicious material from servers if the copyright holder alerts them to the problems. This lawsuit—in which Verizon prevailed—was an attempt to broaden the scope of the law and put pressure on service providers to police their own customers.[25]

The fourth front in the music war was propaganda. In conjunction with publishers such as Scholastic, which publishes the Harry Potter series, the recording industry is trying to convince fifth graders that sharing music is wrong. There are similar efforts to reeducate high school and college-age people about the problems generated by file sharing. These industries hope to alter the cultural norms that govern such unmediated communication and cultural sharing—norms that I argue have been present, albeit invisible and unarticulated, for centuries.[26]

The fifth front was to offer a substitute service, the first element of what was supposed to be that perfectly sealed global jukebox. In 2001 two services started offering authorized selections of encrypted or watermarked music files. MusicNet is a joint venture of AOL Time Warner, Bertelsmann, and EMI. PressPlay is a partnership between the other two major labels, Vivendi Universal and Sony. Before rolling out these two services, which will not offer competitors' songs, the five big companies managed to sue, buy, or starve many upstart firms that had hoped to offer licensed commercial music. Neither service allows users as much freedom and flexibility with music files as MP3 does, but both offer some special services and rates that allow users to burn songs on blank compact discs. Both charge nominal rates per song download. Because these five companies have leveraged their copyrights to dominate legitimate distribution over the Internet, and thus control 80 percent of the commercially available music, the U.S. Justice Department started an investigation into the possibility of an antitrust violation.

Although digital rights management is an important part of these business models, the RIAA no longer pretends that encryption and watermarking will stop the flow of unprotected content over the Internet. Its more modest goal is to reinstall "friction" or "inconvenience" into the system so that everyday nonspecialists will have less incentive to traffic in open and unauthorized content. Yet friction-free music remains free and easy. More than with compact discs, these commercial services compete directly with free peer-to-peer networks. Then in early 2003 the Apple Music Store debuted, offering albums for about $10 and individual songs for

$0.99. The music files came lightly protected by easily defeated digital rights management. For instance, only ten copies could be burned from a single playlist. But ripping a CD made from that playlist made it possible to continue burning CDs from the resultant MP3 files.[27]

In early 2004 it's clear that unregulated peer-to-peer music sharing is not destined to be as free and easy as it once was. Millions still do it. Millions more might try it soon. But they are likely to calculate the costs and risks of such behavior. The RIAA opted out of frustration to seek resolution through the civil courts. This is how copyright law is supposed to work. The short-term result of these suits is that many fans might settle for good services like the Apple Music Store or even for more restrictive options like the newly released Napster 2.0, which offers streams of music from a central service. But the industry seems baffled about how to alter consumer behavior short of suing their best customers.

Then there is the nuclear option: Federal criminal law could be brought to bear in this war. Back in 1994, before relatively anonymous indexing systems like Napster and Kazaa emerged, a Massachusetts Institute of Technology student named David LaMacchia was charged with federal wire fraud for creating a file-swapping system on the Internet. A federal judge dismissed the criminal charges, saying that it was not clear that Congress had intended to criminalize the everyday acts of what was then thousands—now tens of millions—of consumers. Technology journalist Declan McCullagh discovered in early 2003 that a relatively obscure federal law, the No Electronic Theft Act of 1997, was passed to close that loophole and express the will of Congress to prosecute file sharers. The act, which could subject downloaders to a penalty of up to $250,000 and a prison term of as long as three years, explicitly prohibits the "reproduction or distribution" of copyrighted material via electronic networks. Several U.S. representatives have pressured Attorney General Ashcroft to prosecute peer-to-peer file sharers under this act. For now, the U.S. Justice Department seems largely uninterested in prosecuting people like me.[28]

A Rational Revolt?

Soon after I discovered Napster in the fall of 1999, my employer, New York University, blocked access to it from any computer connected through the university's network. This policy was imposed without warning or discussion among the faculty. I protested to the university technology office, to no avail. Clearly those who made this decision had not considered its effect on legitimate research as well as illegitimate copying. The action meant that I could not use my office computer to observe how Napster worked or correspond with the company's staff through its Web site. The first lesson this experience taught me was that the battle over peer-to-peer involved values beyond commerce and crime, including free speech, privacy, and intellectual freedom. On the afternoon the administration sent the e-mail announcing the Napster block I overheard several conversations among students. Names such as Aimster, Scour, and Gnutella flew around elevators and hallways. My inability to use Napster drove me (and several thousand NYU students) to alternative peer-to-peer services then sprouting up. This brief experience taught me the second early lesson about peer-to-peer: an attempt to quash one service was likely to backfire on those hoping to discourage unauthorized downloading.

These two lessons worked their way into a rather preachy and simplistic article I wrote for *The Nation* in July 2000. In the article I described the success of Napster as "a rational revolt of passionate fans." I argued that Napster users like myself were rebelling against the inflexible price structure, limited consumer choices, irresponsible gatekeeping, and technological oligopolies of the music industry. Music consumers, I claimed, would soon be making their choices with the benefit of better information. Naively I asserted that Napster builds community, much as the punk rock and hip-hop movements did. My great hope was that regardless of the direct effect on CD sales, MP3 distribution would make music fans more informed consumers and music companies more responsive producers, better serving the margins of the market, such as ethnic communities, subcultures, and political movements.[29]

I still agree with some parts of the article. For instance, that Napster as a company and legal target was a grand distraction; that the real story was the convenience of MP3 compression and the genius of fully distributed systems like Gnutella. But I was wrong about peer-to-peer building community. Peer-to-peer networks can serve communities and subcultures of fans that already exist in cyberspace or the real world. But they don't do much to build new communities where none exist. Few peer-to-peer systems allow fans to deliberate about music or make connections and decisions collaboratively. Peer-to-peer systems work as the simplest and most frustrating of libraries: Each service offers an efficient yet dumb textual search index. Often you find what you already know about. But they don't help you move beyond it; there is little opportunity for serendipity. In this way, the anarchy embodied in peer-to-peer file sharing is somewhat childish, simplistic, and demand-driven. The users who can scour the hard drives of others and download enough material to become creators and mavens in the community tend to be wealthy, technically savvy residents of developed nations. And even they have few ways of building social capital or deriving cultural capital directly from peer-to-peer.

Nobody knows or rewards the blues fan who is nice enough to offer a copy of Son House's rare classic, "My Black Mama." Yet people do that sort of thing. Everyone seems to have a grasp of what motivates a person to download music. But what prompts a person to offer thousands of songs—many obscure, rare, funny, or remixed? Peer-to-peer libraries of music only offer us substance to enhance cultural relationships we already have in real life. *The Nation* article allowed me to address the conflicts and questions inherent in the peer-to-peer story. But it did not offer the chance to ask the really important questions about this communicative technology and its cultural effects. For that, I needed a good argument.

Enter the Ethicist

Let's face it. I am not the typical user of peer-to-peer music services. I have an agenda. I subscribe to an ideology or three. I read and write about electronic distribution systems more than I actually use them. I am an activist as well as a scholar of media systems and cultural politics. I want to keep these channels open and the music flowing, and I want to change the terms of the conversation that has dominated the accounts of the phenomenon. So I boast of my good behaviors—like my excessive compact disc purchases—to justify my behavior. Yet my behavior is ethical only on the basis of my utilitarian assumptions: I do not harm—and perhaps help—the commercial music industry in the long term. The net "cash value" of my contributions to the music economy is positive. If you find such an ethical universe unacceptable, and you consider art and property fundamentally inviolable, perhaps sacred, then you should conclude that my behavior is unjustified and my explanation is unpersuasive. After acknowledging that difference we could have a great conversation about ethics.

Most users of peer-to-peer file sharing, however, do not have the luxury of spending many hours contemplating and debating the ethical dimensions of free-flowing culture. Although some might behave ethically anyway, many do not. Let's face it, plenty of people are compiling entire Eminem albums from peer-to-peer systems and opting out of the commercial music economy (it's not as easy as some claim). Although clearly, as Eminem's success testifies, many more are still willing to shell out $16 for his compact discs.

Randy Cohen, who writes the Ethicist column for the *New York Times Magazine*, shares my privileged status. We both have these kinds of conversations for a living. Cohen took on the issue of peer-to-peer use in the summer of 2000. A college student from New Jersey wrote and asked whether his actions were ethical. The student had been grabbing songs that he wanted to own yet did not want badly enough to justify buying an entire album. Cohen, alas, seemed unconcerned about the specific nature of the students' actions and derided any and all use of downloaded music. "To download music from the Net illegally is theft, depriving songwriters, performers, and record companies of payment for their work," Cohen wrote in his column. Cohen argued that by demonizing the wealthy record companies, peer-to-peer fans are "coming perilously close to blaming the victim." His response offered the student no gray areas, no caveats. He simply declared that such behavior is illegal and thus unethical.[30]

So I called him on it. Well, I e-mailed him on it. I wrote to Cohen immediately and took issue with his response. I said that appealing to law was not sufficient when considering complicated ethical decisions. Some laws are badly drawn or misapplied, and breaking them under certain conditions is ethical. Of course Cohen understands that. Besides, I wrote, the law is not clear about which actions are copyright violations. On the date of our debate, courts had not been willing to quash entire technologies or declare private, noncommercial copying illegal. Until the record companies sue or prosecute people like me, we won't know which actions are truly "illegal" (at the time I did not know about the No Electronic Theft Act). Although I consider the particular behavior of that New Jersey college student largely unethical—he should plan to buy the album if he likes the songs and he can afford it—I described my use of peer-to-peer and asked him whether he though it was unethical.

Cohen responded that while my unauthorized use of copyrighted material might ultimately benefit my favorite artists, the availability of these files should be up to the artists, not me. So I asked him about the pragmatic calculus. "What is the practical difference between listening to downloaded music in the privacy of my apartment and listening to broadcast music in the privacy of my car?" Cohen again asserted that the difference was in the permission. He said his chief concern in this ethical question was the interest of the artist—both financial and moral. "The history of popular culture is a continuous struggle on the artists' part not to get robbed . . . it seems to be that what MP3 does is democratize the ability to rip off an artist," Cohen wrote to me. "And what's particularly galling is that you not only want to do it, you want to be praised as a social progressive when you do."

He got me. That's my shtick. By the guiding principles Cohen deployed in our peer-to-peer debate, I had no escape. He considered copyright to be an artists' right and concern; I consider the chief player in the copyright system to be the corporation. He considered copyright a moral right. I consider it a state-granted limited monopoly for the purpose of generating and spreading culture and information (and thus further limited in situations in which the monopoly market fails to spread culture and information at a reasonable price). He never mentioned record companies in his arguments. I never considered the passionate interest of songwriters and composers (who hold a wide range of opinions on the practice). By juxtaposing our arguments, though, we both found it harder to cheat. We had to consider the ground on which the other treads.

Since my exchange with Cohen, I have had the luxury of similar conversations with composers, musicians, record industry executives, and hackers. Few of these rather subtle and complicated terms of debate have worked their way into the rhetoric of policymakers in Washington, D.C., or Brussels, Belgium. Most newspaper accounts of peer-to-peer battles have changed from sports or crime stories to business stories. It's not yet a cultural story, an ethical story, or a political story.

In the absence of a methodical, public discussion about the ethics of peer-to-peer, we are left with blunt, clumsy, confusing legal rulings. Mostly we are left with lies, myths, overstatements, predictions, and threats from all sides. There is a very good chance that for the next few years we will get to live with both open, free, peer-to-peer file-sharing systems and the commercial record industry. So we need to start discussing the most ethical ways for all parties to behave. Alas, neither the anarchists nor the oligarchs are willing to facilitate a dialogue that might allow us to get the best out of both models of distribution. In a mediascape dominated by the tendencies and assumptions of anarchy and oligarchy, with a Congress that seems more interested in prosecution than conversation, calling for rich dialogue sounds like the worst sort of mushy liberalism. Yet mushy liberalism may be precisely what we need.

CHAPTER 5

A Work in Progress or the Final Edit?

One of the reasons there was no immediate and widespread Napsterization of commercial films in the first few years of the twenty-first century is that Napster only allowed its users to trade MP3 audio files. The recording industry may have been encouraged by its court victory over Napster and embarrassed about threatening computer scientists. But soon it faced more formidable foes: two competing systems each achieved the more powerful form of a distributed peer-to-peer system. Both the commercial network Kazaa and the voluntarily built system Gnutella are fully and globally distributed. Gnutella is virtually immune to legal assault. No one owns or runs Gnutella. To shut it down, media companies would have to disable network access for hundreds of thousands of users. Complicating matters more, Kazaa literally is a distributed software company, as well as a company that distributes software that distributes software. Its investors have learned from their own customers and design principles. As Amy Harmon of the *New York Times* describes the company, "Sharman networks, the distributor of the program, is incorporated in the South Pacific Island nation of Vanuatu and managed from Australia. Its computer servers are in Denmark and the source code for its software was last seen in Estonia."[1]

After the demise of Napster, the film industry watched users flock to networks that enabled them to trade any kind of digital file: music, video, photos, or software. Video anarchy suddenly mattered for reasons that went far beyond the file-sharing habits of technically adept American consumers. Video anarchy threatened to undermine both the market for commercial films and the status of professional filmmakers and their financiers. So the Motion Picture Association of America and its well-connected and widely respected president, Jack Valenti, stepped forward to set the tone, agenda, legal precedents, and vocabulary for the next set of conflicts over the control of culture and information. The motion picture industry extended this battle from the Internet into the devices and habits that make up the domestic mediascapes of millions of people: televisions, video recorders, home computers, and the editorial control that users may exert over the images that came into their lives.

Immediate Gratification

If I want to watch *Harry Potter and the Sorcerer's Stone* on video this evening, I have several options. I can walk down the street to Kim's Video Store and rent the VHS tape or DVD for about $3. Or I can purchase the DVD for about $18. Or I can walk a few blocks south to Canal Street in Chinatown and buy a pirated VCD of it for about $4. Or I can log on to a peer-to-peer system this morning and use my high-speed cable modem to download a digital video file in two segments. The first segment takes about four hours to download, even at the highest speed Time Warner Cable allows me to use. The second segment takes about two and half hours. What's my best option?

Fortunately for Hollywood, my best option is the legal and legitimate one. Renting a DVD from a video store is cheaper, easier, more dependable, and more convenient than the other alternatives. Millions of Americans have made the same choice. That's why the DVD player was adopted more quickly than any other media technology in history (with the possible exception of Napster). It's also why, even as commercial music suffers from CD sales drops of between 6 and 10 percent per year since 2000, the film industry is doing better than ever.

If high-speed broadband gets faster, Hollywood might have something to worry about: Greater numbers of quality films might show up on Gnutella as stable, complete files. Right now I am as likely to get a falsely labeled porno film as the real Harry Potter movie from Gnutella. And Hollywood is not above creating unjustified "moral panics" to generate support for expansive new technological systems designed to enforce control over its products and enable more efficient, profitable ventures. Still, as of spring 2004, Hollywood is riding high.[2]

Real piracy, rampant on the streets of major cities around the world, poses a threat for Hollywood. For-profit ventures are run by organized crime syndicates or small independent entrepreneurs who have invested a few thousand dollars in machines to generate unauthorized copies of commercial films. Pirates acquire their content by sneaking digital video cameras into theaters showing first-run films in the United States, by stealing or borrowing reviewers' VHS tapes, by copying store rental tapes, and by hacking the access controls around DVDs. Then they make bootleg copies, mostly of poor quality (often with "do not copy" warning tags running along the bottom of snow-flaked screens). These are not hobbyists, hackers, artists, or collectors.

Real piracy is not like the private, noncommercial copying shared among friends that characterizes American-style leaks in the copyright system. Still, it's impossible to measure the real effects of this anarchistic black market because—just as with more benign noncommercial peer-to-peer distribution—you can't assume a one-to-one correspondence between pirated copies sold and legitimate sales lost. Poor quality copies released early might generate a desire to see the better quality version in authorized form later, upon theatrical release. Piracy thus serves as unintended advertising. Often pirates bring films to places in the world where they would not otherwise open. In countries with conservative social mores, for instance, the theatrical release of a film might be heavily edited to remove the naughty bits or politically controversial expressions. The pirated version available on the streets might be the only American release version. Nevertheless, it's not hard to assume that widespread video piracy undermines the market for legitimately released commercial films.[3]

Adding to Hollywood's angst, workers in the studio system, such as directors, writers, and editors, feel threatened by digital technologies that empower consumers to become editors of major Hollywood films. This is a moral threat, not an economic one. But it shows that these new technologies generate tensions and affect how professionals see themselves in the world, exposing uncomfortable relationships with consumers.[4]

Hollywood's potential problems with technology differ from the travails of the music industry. Peer-to-peer networks as we currently understand them, despite Jack Valenti's protestation, don't directly threaten Hollywood's markets for reasons that have to do with the complexity of the content and the social practices of movie fans. People enjoy theaters and movie rentals are convenient and cheap. Someday consumers may tire of the theater experience and the price of popcorn or perhaps high-speed data pipes and compression formats won't improve significantly. But we have no way of predicting such changes. Despite our familiar habits of techno-fundamentalism, we have no grounds to predict the rise of private, peer-to-peer digital leakage of major commercial films.[5]

Informal economies, black markets, and other anarchistic distribution networks in the real world challenge Hollywood, as well as commercial film industries around the world. These networks are more like electronic gossip than a service such as Napster. Fans and culture groups trade information and criticism about films, which is important for the distribution, reproduction, and consumption of films that do not have huge global publicity machines or a star system behind them. For diaspora communities like Han Chinese or South Asians, these networks are of major importance.

Consider my options if I want to watch the Hindi movie *Asoka*. Hindi, Bengali, Malayalam, Telegu, and Tamil films are among the bundle of strategies that members of the South Asian diaspora use to strengthen their cultural identities. The songs, dances, and characters of Bollywood and its subcontinental satellites help Indians maintain and extend a postmodern sense of being Indian without being in India. The Mexican, Chinese, Turkish, and Nigerian diasporas, as well as the transnational South Asian communities from Singapore to Sidney to Los Angeles to London, share these portable elements of culture. As it turns out, Kim's Video Store in my neighborhood does not carry *Asoka*, which is a poor dramatization of the life of the first Buddhist king of northern India. He introduced and enforced religious freedom in ways that European Enlightenment thinkers would later profess. I called around to some of the dozens of Indian video stores in the New York City area. Some have pirated copies of it. Some have legitimate copies of it on VHS or DVD or VCD. I found the title on Gnutella, but I was averse to the long download time, heavy pixelization of the picture, poor synchronization between video and audio, and sitting at my office desk for more than two hours to watch the video. I considered ordering the film and have it delivered through the mail. It's available on VHS and DVD from Amazon.com, but it would take several days to arrive. From other on-line sources, the tape or disc that arrives might be legitimate, or it might be pirated. If I seek immediate gratification—the promise of the global jukebox—I am likely to hit the black market whether I intend to or not.

Despite the more immediate threat posed by real piracy, the Motion Picture Association of America has focused its high-profile efforts against hackers, peer-to-peer electronic networks, home recording, and the very concept of the personal computer in the United States and Western Europe—which it sees as the next front in the battle to control the flow of its content. Importantly, it is attacking those who revise, reuse, and reedit video to create more palatable experiences or more interesting art. More than piracy, Hollywood is reining in creativity, adaptability, and customizability.

Hollywood Versus Hackers

Since its introduction in 1997, the DVD player has reached more than 46 million American homes and 3 million in the United Kingdom.[6] The movie industry claims it never would have introduced the DVD format if the U.S. Congress had not passed the 1998 Digital Millennium Copyright Act (DMCA), which gave it confidence to install an access control program called the content scrambling system, or CSS. The problem with this claim—often offered as a defense of the DMCA—is that Hollywood introduced the DVD four years before Congress passed the DMCA. Faith in its digital rights management system was central to justifying investment in the new format. CSS is a series of encryption algorithms that authorize a player to access the data on a DVD. It acts as a handshake between the player and the disc. In this way, CSS prevents unauthorized machines from accessing the content.

The movie industry entered into licensing deals with DVD player manufacturers to ensure stability, quality, and standardization in the industry. A player that lacks permission to unlock CSS cannot participate in the DVD market. Manufacturers use the CSS consortium to limit the variety of players on the market, and the studios ensure that DVD players lacked certain features such as digital video outputs or hard drives that might allow for home copying, archiving, and editing of DVD content. Back in the early 1990s, as CSS and the DVD were in development and the DMCA was just a U.S. Commerce Department white paper, film companies had already imagined the regulatory nightmare of peer-to-peer electronic file sharing. Yet even after the rise of Napster, it was unclear how peer-to-peer networks would undermine the DVD market. As peer-to-peer has grown, so have DVD sales.

To convince the public that the Napsterization of video was a likely prospect if the integrity of CSS was undermined, the industry conjured two myths. First, it had to prove that peer-to-peer systems made it easy to acquire good copies of digital video. Second, it had to show that any algorithm or program that evaded, defeated, or circumvented CSS was to blame for such widespread availability. It would be hard, however, to show that hacking through CSS contributes to piracy or peer-to-peer distribution for one simple yet often ignored reason: CSS regulates access and compatibility, not copying. Anyone can copy the data on a DVD with little effort. Playing a DVD on an unauthorized machine is another matter.[7]

In 2000 a group of hackers generated a small program they called DeCSS, which, as the name implies, decrypts CSS. The first version was about thirty lines of code, although later versions included as few as eight lines. The mere existence and the simplicity of DeCSS showed the computer security world how ridiculously lousy CSS was at protecting its content. These hackers produced DeCSS for a simple reason. They used computers that run the open-source Linux operating system, and CSS was only licensed for use by Microsoft Windows and Apple Macintosh computers. To play their DVDs on their laptops, they had to evade the access control device. The existence and deployment of DeCSS had no significance for those who distribute pirated DVDs. Yet the MPAA insisted that there was a connection, and it went to court to make a firm statement against those who used and shared the decryption code. Though the MPAA argument was largely spurious, DeCSS was a threat because it revealed the futility of the movie industry digital rights management strategy.

Emmanuel Goldstein, the nom-de-keyboard of the editor of the influential hacker magazine *2600*, found himself in trouble almost immediately. His Web site had run some articles that explained the DeCSS code and included the code in the text. At the request of the movie industry, a federal judge issued an injunction prohibiting the journalist from describing or revealing this phenomenon. Forbidden to run the code, Goldstein put up hyperlinks on the *2600* Web site that pointed readers to other sites that contained the illicit code. The judge then issued an injunction against the hyperlinks. This was a wonderful case for the movie industry. The defendants were assumed to be long-haired, ill-mannered, antisocial, mysterious hackers. After a decades-long fight for its own free speech rights, the film industry seemed blasé about quashing the rights of hacker journalists. The movie industry prevailed in federal district court, without ever showing that DeCSS or any of the defendants had actually contributed to the infringement of any copyright or the distribution of any video. The plaintiffs merely had to show that the code decrypted CSS and was thus a device intended to circumvent an access control system. A federal appeals court took up the First Amendment questions about whether computer code is speech and thus protected from most government regulation, and whether a court could regulate hyperlinks on the World Wide Web. On both issues, it agreed with the movie industry and the U.S. Justice Department and against the hackers, journalists, computer scientists, civil libertarians, and academics who lined up to defend the speech rights of the magazine. The court ruled that Congress had a compelling interest in preventing piracy (even though CSS does not prevent piracy, DeCSS does not enable it, and no piracy was alleged), which trumped any claim to the free flow of code or the ability of Web site authors to link to whatever sites they wish.

Strangely, the motion picture industry spent many months and untold dollars trying to stem the flow of eight lines of PERL script computer code—the digital equivalent of firing artillery shells at a cloud of gnats. The industry fought the lightest, most portable, most discreet, and most translatable of infringements. Wouldn't it be easier to stop the distribution of several megabytes of data flowing through a pipe than eight lines appended to an e-mail? Wouldn't it be more productive to spend money stopping sidewalk DVD vendors in New York than hackers who want to discuss the intricacies of encryption? Yet the sidewalk vendors remain in business, and *2600* remains enjoined. Meanwhile, the code quickly went underground—distributed via peer-to-peer systems like Gnutella—and out in the open. Within days of the injunction, T-shirts appeared with the code emblazoned on them (with the headline, "I am a circumvention device"). People wrote poems and songs that expressed the code lyrically. Internet users appended the code to the signature sections of e-mails, thus making everyone who forwarded these e-mails a violator of federal law. I carry the code around in my handheld computer, ready to beam it to anyone who wants to violate federal law with me and show how futile and ridiculous the law is. Yet the MPAA claimed a victory. It won in court in 2001 just as it had won in Congress in 1998. But it lost the larger struggle to maintain control of its digital content.

The Folly of Digital Television

You can get almost anything you want from the U.S. government if you promise to deliver digital high-definition television to the American people. Dreaming of high-definition video and stunning sound from our free airwaves, we have already been suckered into giving up $70 billion worth of broadcast spectrum for the dream of really cool video, while receiving nothing in return from broadcasters. Hollywood is holding its prime content hostage until we give it what it wants—control over how, when, how many times, and on which machines we record and play programs in the privacy of our homes. In 1996 the broadcast industry said it would offer consumers high-quality, multichannel programming over terrestrial signals if the government would grant it $70 billion of spectrum for free. But the broadcasters had to promise to give back their old spectrum once they got the new signals flowing. Congress fell for the scam, giving broadcasters everything they wanted and exacting nothing in return; we are still waiting for our spectrum dividend. It was one of the grandest examples of corporate welfare in American history. Yet, having learned nothing from the debacle, the federal government is again rushing in to protect and enrich established industries at the people's expense.[8]

At first glance the folly of digital broadcast television seems to be a classic case of market failure. Firms and industries had to make substantial investments and take some risks to roll out digital broadcast television. Yet no single actor, fearing the other players would fail to come through, wanted to make the first move. So the government decided to encourage the various players to make the necessary moves. Since no one even asked the American people if they wanted digital broadcast television, this was a failure without a market. Only 7 percent of American viewers get their television programming through rooftop antennas; the rest use cable or satellite systems that already offer digital quality. And of rooftop antenna users, only 10 percent have been interested in spending $250 or so to get a special digital tuner to receive the new digital broadcasts. Meanwhile, broadcasters are less and less interested in using their valuable new spectrum for high-definition television signals, and more interested in exploring ways to roll out other ancillary digital service such as multiple versions of their current channels or wireless Internet service. Since broadcasters won't offer their standard fare over digital spectrum by 2006, they won't have to give back their old spectrum so the government can auction it off again. They don't have to give anything back until 85 percent of the televisions in their local markets are capable of receiving digital broadcast signals. As of April 2003, fewer than 16 percent of broadcasters were even transmitting digital signals.

Surprisingly, the networks have played along with the farce, offering much of their prime-time programming and sports in high-definition formats, even though consumers don't want to pay the premium for it and most broadcasters don't want to send it through their channels. The Federal Communications Commission, which is more than happy to invoke laissez-faire principles when it comes to competition in the industries it regulates, is now forcing electronics manufacturers to install digital broadcast receivers in new televisions, which could add hundreds of dollars to the cost of each set and only benefit the 7 percent of Americans who use broadcast signals.

After watching broadcasters milk the public over digital television for eight years, Hollywood wants something too. It promises to offer major motion pictures in high-quality digital formats to broadcast stations in hopes that regulators will keep digging the regulatory pit deeper. The big studios want to control what and how you record and distribute television programs in your home. They claim they need to rein in the rampant sharing of digital files over peer-to-peer networks, even though sharing will continue as long as hackable DVDs exist. To gain this remarkable control over our personal mediascapes, the major studios have proposed "broadcast flag" standards to guide the use of digital signals.

These new FCC standards require televisions to include electronic devices that regulate accessing and copying encrypted material. They would determine which devices could and could not play material that contains small bit of code called a "broadcast flag." The presence of the flag would tell a digital device (computer, home digital recorder, television, etc.)

whether the content is authorized to be played. The regulations would allow you to record *The Sopranos* for later home viewing on an old machine such as a VHS recorder. But VHS is a dying medium. Many electronics stores are dropping VHS players from shelves in favor of digital video recorders like TiVo and nonrecording devices like DVD players. Video stores are expanding their DVD offerings and shrinking VHS space. As they break down, VHS machines will find their way into landfills and museums, taking substantial consumer autonomy with them. There are two major problems with the broadcast flag. The first has to do with consumer rights. The second—and perhaps more important in the long run—concerns creativity and innovation. One of the basic tenets of media law in the past two decades has been that users have a certain amount of autonomy in deciding how, when, and in what form they use lawfully acquired content in their lives. Private, noncommercial, noneducational uses are generally considered noninfringing. The fair use concept has grown into a penumbra of rights that copyright users confidently enjoy without fear of being sued. Either the law does not explicitly forbid such uses (such as making mixed tapes or CDs), or it explicitly allows them (such as time shifting television programs).

A slim 5–4 Supreme Court decision in 1984 preserved the ability to record shows at home for later viewing. Back when Jack Valenti of the Motion Picture Association of America was crying that the VCR would destroy the American film industry, the court considered whether a copyright holder's rights to limit all copying extended to the living room and whether the VCR had "substantial noninfringing uses" that would mitigate calls to outlaw it. As Justice John Paul Stevens wrote at the time, "to the extent that time-shifting expands public access to freely broadcast television programs, it yields social benefits." Since then, home recording and "time shifting" (recording a program for later viewing) have been considered fair use. Additionally, many people have taken to "space shifting" copyrighted materials onto peer-to-peer networks, creating libraries of low-quality digital video that worry Hollywood. The only way to stop this practice is to make it technologically inconvenient or impossible to move content from one medium to another. The broadcast flag proposal would not shut down analog home recording but would prevent the redistribution of flagged digital content. Other efforts by the MPAA would close the "analog hole" through digital watermarks that would prevent machines such as your home computer from converting analog to digital signals. Many people now record analog programs in digital form through their personal computers. They take bits and pieces of shows and films and splice them up into parodies, pastiches, or completely new creations. They make archives of their favorite *Simpsons* episodes on discs. But if Hollywood (through the FCC) can mandate digital controls in hardware and then rolls out content according to its own standards, these practices could end. The technically savvy pirates of the world would still be able to traffic in illicit materials. But the rest of us would be unable use materials in the way we have become accustomed. Users' rights would be limited—not by changes in the fair use law but by technological modifications that make it difficult to exercise without explicit permission, and limit both technological and artistic creativity.

Government-mandated technical standards would lock certain firms into the lead in the digital television technology market. Innovators on the sidelines, whether entrepreneurial small engineering companies or garage tinkerers, would be limited in the sorts of software and hardware they could deploy for future digital television devices. They could not build and market their own digital receiver, no matter how smart and creative they are, without agreeing to the standards and guidelines that Hollywood would dictate and the federal government would enforce. Perhaps worse, the government might be locking in bad technology. The requirement of the digital television broadcast flag is an example of clumsy technological, legal, and cultural policy. It's a grand overreaction to an undemonstrated problem in the service of a specious policy goal. The rule-making process charged the commission with considering the effects of this proposal on users' rights. But the commission failed to consider whether it is in the nation's long-term interest to lock in particular technologies and protect certain powerful industry players at the expense of innovation. As technology and legal expert Mike Godwin has written, "There's more than digital television at stake. Bad government actions in this sphere—and you can be sure that Congress and the Federal Communications Commission will act rather than refrain from acting—could permanently shoehorn part or all of the computer revolution under government-driven design control. Not only would this likely kill the dynamism of the information technology sector, but it is unlikely to do much to protect copyright interests."[9]

Just Another Appliance

Reengineering the television and home video recorder is only the beginning of the MPAA agenda. In 2002 the MPAA worked through Senator Ernest Hollings of South Carolina to introduce the Consumer Broadband and Digital Television Promotion Act. This bill, which has not been approved by Congress, would require all digital hardware manufacturers to embed "security technologies" in their machines. The list of machines affected by this proposal is virtually unlimited—CD and DVD players, all personal computers, televisions, digital personal video recorders. Edward Felten has concluded that the requirements would also cover musical car horns, microwave ovens, some refrigerators, and perhaps a toaster oven or two. Any product that employs digital technology or interprets digital signals would be covered by the proposal. But the real problem, as law professor Pamela Samuelson declares, is that the proposal "aims to outlaw the general purpose computer." Desktop personal computers do four things: store data, retrieve data, compute, and copy. All the video games and Internet communication we observe are simulations generated by the complex, high-speed combination of these activities. The very act of storing and retrieving data involves copying. That's basic and essential to a computer's design. By mandating that all new computers include copy control technology, the government would be shifting control of these machines to the content holders and limiting the power of the computer owner and user. Home computers would be governed by remote control. What's more, this proposal could render computers that run the Linux operating system illegal, because openness in copy control technology would be counterproductive, and Linux, like the larger design of both the personal computer and the Internet, demands a certain level of openness.[10]

The Final Edit

If Hollywood studios could deliver their dream products in their dream formats, they would send every first-run film to thousands of theaters around the world via electronic pipes. Digital projectors would emit high-quality images on screens. The studios would control which versions got to which theaters. Theaters in Saudi Arabia, Pakistan, India, Singapore, or

Utah might receive versions without nudity. Theaters in New York, Amsterdam, and San Francisco might receive versions with extra nudity. If audiences registered disappointment with a particular ending, studios could beam out a revised version with a new ending. Studios could even send multiple versions to the same theater—a PG-rated version for all shows before 8:00 P.M. and an R-rated version for all shows after 8:00 P.M. The storage capacity of DVDs would allow multiple versions on the same disc, so that families could watch *Titanic* without the nudity if the kids were in the room and with it when the kids fall asleep. With each home connected to a pay-per-view jukebox, there would be no need for the DVD. Families could just order up their preferred digital stream. Ideally, of course, Hollywood would save on the cost of casting and reshooting scenes by replacing as many human beings (or "blood actors" as they are known) as possible with computer-generated cartoons.

There are formidable obstacles to this dramatically efficient vision, first and foremost, the cost. No one wants to pay billions of dollars to retrofit theaters with digital projectors. Until there are enough digital projectors, there is no incentive to distribute digital prints. Human beings themselves are formidable obstacles. Actors, directors, and editors, who have some power in Hollywood, do not want their labor replaced or their status as artists compromised. Studios already issue different cuts of films for different foreign markets and airline viewing, after negotiations with directors and editors, and after the films have either failed or succeeded in domestic release.

As Hollywood creeps toward this digital vision, George Lucas leads the pack. His last two films, *Star Wars Episode I: The Phantom Menace* and *Star Wars Episode II: Attack of the Clones,* were filled with digitally generated extras where blood actors might have served in the 1970s. Several major characters, including the inexplicable Jaba the Hutt and the blatantly racist and annoying Jar Jar Binks were (fortunately) digital creations. The same technology that allowed Lucas remarkable control over his characters gave his fans the opportunity to undermine his control of them. Early in 2001 rumors began flying around Internet sites and chat rooms that someone had taken *Episode I: The Phantom Menace* and created something called *Episode I: The Phantom Edit.* The phantom editor, who remains incognito, shortened the film by about twenty minutes, removing most of the scenes that focused on Jar Jar Binks. Without dialogue, Jar Jar Binks was much less offensive. The phantom editor also removed some of the stilted dialogue and awkward verbal gestures that Lucas had installed to appeal to children. Soon after the rumors of the edit started spreading, copies began appearing in VHS form at Star Wars and science fiction conventions. Digital copies flew from peer to peer and via peer-to-peer networks like Gnutella. The seven-hundred-megabyte DivX file took many hours to download even with the fastest connection available. But the demand for the file was not about getting *The Phantom Menace* for free. It was about seeing a better version and celebrating the anarchistic revolution that allowed a lone film critic to take control of the content and connect with thousands of others who shared his appreciation of the Star Wars saga. Lucas was reportedly curious about the cut, but his company, Lucasfilm, warned fans that sharing these copies and files constituted copyright infringements of the original film.[11]

Other directors were not amused by the technological power available to those who are not part of the Hollywood system. In late 2002 the Directors Guild joined the major studios in a lawsuit against a Denver-based company called CleanFlicks, which edits potentially offensive material from Hollywood videos. These "family-friendly" edits satisfy a market of religious and conservative families that Hollywood has not served very well. Two issues lie at the heart of this conflict. First, American copyright law recognizes that copyright holders—in this case, the studios—control the right to create "derivative works" of their holdings. Second, there is the right of a creator to control the reputation and integrity of her works—the directors' appeal to their "moral rights." Such rights are not central to American copyright law, which appreciates the process of revision and play with older materials (and the power of corporations to have the ultimate authority over content), but they are strong in French and Continental artistic law.[12]

Recent cultural theory emphasizes the "meaning making" power of audiences. Because the producer of a work cannot know or dictate how a particular community of readers, listeners, or viewers will make use of a cultural text, the producers' intentions are far less important than the ways audiences make use of the messages and images. The television show *Dallas* might do very different things in the minds and lives of people in the United States versus those in China or Egypt. In the United States, Homer Simpson is a unique character, interesting for how he deviates from American behavioral norms. In Europe, he is a typical American—fat, rude, stupid, and provincial. This approach to culture is controversial, of course. China and Egypt are no more culturally monolithic than the United States is. And young people might read things differently than old people. Rich people see different things than poor people. Even within the same family people take and use meanings differently. Every film is *Rashomon.*

Imagine if we could go beyond exercising control of our individual critical faculties. Suppose, in addition to reading things differently, we could rewrite them. Imagine if we could make the most powerful images in our world more to our liking, more relevant to our lives. Would this produce a radical change in our mediascape and consciousness? Until the rise of fixed and legally protected media products like television shows and feature films, humans had the power to adapt and reuse cultural elements. American communities quickly adopted Harriett Beecher Stowe's novel *Uncle Tom's Cabin* to the local stage and undermined its abolitionist messages. Uncle Tom was soon a stock comic character in minstrel shows. Stowe gave birth to Uncle Tom but America kidnapped him, changing him into something she would neither recognize nor celebrate. Those are the risks of releasing messages into the world. An author cannot control how a character, idea, or plot will be read, refashioned, or criticized. But the restrictions that copyright law places on the production of derivative works and the integrity of the original work alter that dynamic.

Technological barriers also limit what audiences can do to material. But the phantom edit shows that this barrier is crumbling quickly. Perhaps the most extreme case of pirate editing was the goblin edit, by an amateur Russian digital video editor named Dmitri Puchkov. Not satisfied with merely watching illegal copies of Hollywood films, he differentiated some products in the rather crowded Russian video market. The goblin redubs films into colloquial Russian, trumping the rather unsatisfying subtitle translations.

The goblin's greatest hits are the redubs of the first two parts of the *Lord of the Rings* trilogy. He turned Frodo Baggins into Frodo Sumkin and the rest of the "good" characters into caricatures of incompetent Russian officials. The evil Orcs became Russian gangsters. Gandalf the Wizard constantly quotes Karl Marx. Puchkov originally made the new versions for his friends, but they made copies and spread them widely. Pirate video merchants all over Russia are distributing goblin edits, which are in high demand, for about $10. The goblin is currently working on a Russian *Star*

Wars edit. By throwing out the old soundtrack and revising the characters completely, the goblin is producing a fairly new work that does not directly compete with the original in the marketplace. No one interested in the original "good" Frodo Baggins would take the goblin version instead. But the real value of the goblin edit is that it uses an English text and Hollywood production (as well as New Zealand settings) to comment on Russian politics and society. This is multilayered cultural criticism and revision on a par with the minstrelization of *Uncle Tom's Cabin*, Woody Allen's Occidentalization in *What's Up Tiger Lily?*, and Leonard Bernstein's urbanization of *Romeo and Juliet* into *West Side Story*. It will make some feel queasy and others giggle. It should make everyone pause and think.[13]

If all films are considered permanent "works in progress," what does this imply for the status of Hollywood labor, as writer Peter Rojas has asked? Should creativity be reserved for professionals and experts? Or will teenagers in their basements and libraries be able to soup up or strip down the signs, symbols, and texts that make up such an important part of their lives? Will Hollywood, bolstered by the political power of the United States government, be allowed to dictate the form and format of distribution around the globe? If powerful media companies use law and technology to ossify their advantages, how will this affect local cultural forms? In lawsuits, congressional hearings, and international negotiations, Hollywood studios claim they need maximum and nearly permanent control over their products to justify their massive investments in production, marketing, and distribution. Although the issue is cultural as well as commercial, the film industry and the governments that do its bidding are willing to go to extreme measures to preserve their global cultural and commercial standing.

CHAPTER 6

Imagineering

To Control the Culture Is to Control the Future

Alice Randall grew up in the United States of America, a nation that defines itself through a handful of ideas, philosophies, and central texts. Among these texts is the 1936 novel *Gone with the Wind*. Margaret Mitchell, a white woman from Georgia, composed this grand love story of defeat and redemption at a time when most Americans believed that the Civil War of 1861–1865 was a tragedy, an operatic event that forged the national character at the expense of southern gentility and tradition. Randall, an African American woman, had deep affection for (and growing misgivings about) the myth of the South perpetuated by Mitchell's novel, and she believed there was more to say about the issue.[1]

Consequently Randall composed a revision—an addition, an addendum—to the classic novel that had burned its way into American mythic consciousness. She generated a manuscript she called "The Wind Done Gone." She sold the rights to publish it to Houghton Mifflin, a distinguished publishing house. As the publication date neared, SunTrust Bank, which runs Margaret Mitchell's estate, filed a lawsuit against Houghton Mifflin, claiming that the new novel was a "derivative work" that depended on the plot, characters, and success of *Gone with the Wind*.

As Lawrence Lessig points out, when Mitchell published *Gone with the Wind* in 1936, the maximum copyright term she enjoyed was fifty-six years. But eleven times over the past forty years Congress has retroactively extended copyright terms for works already published, in 1998 adding twenty years to almost all copyrights then in effect. *Gone with the Wind* should have entered the public domain in 1993. The story and its characters long ago entered the pantheon of American texts that have informed the nation's identity. But the nation had yet to assume ownership of the story.[2]

The Mitchell estate had already commissioned a sequel to *Gone with the Wind*, the novel *Scarlett*, written by Alexandra Ripley and published in 1992. A critical embarrassment, *Scarlett* was a best-selling book, certifying the mythic power that the original characters had in the American imagination, as well as their commercial potential. Clearly the estate considered ownership and control of the plot and characters worth defending, as well as the reputation of the fictional characters. *The Wind Done Gone* took liberties with the behavior of the residents of Tara and Atlanta; chivalry and honor did not play a central role.

When the two sides met in an Atlanta courtroom, the lawyers for Houghton Mifflin argued that *The Wind Done Gone* is a parody of *Gone with the Wind*, a point that could potentially stop the lawsuit in its tracks. The U.S. Supreme Court had ruled unequivocally in 1991 that elements of an original copyrighted work were allowed as fair use in a parody. Comic genius Carol Burnett had spoofed *Gone with the Wind* in her classic sketch show of the 1970s. Why shouldn't Alice Randall enjoy the same artistic freedom as Carol Burnett?[3]

Parody is one of the few aspects of the tangled, vague principle of "fair use" that has emerged in the twenty-five years since the U.S. Congress codified it in federal statutes. Fair use is harder to employ, harder to define, and harder to explain now than ever before. Artists, scholars, and writers all over the United States avoid quoting passages from books or songs for fear of being sued. As a result, borrowing to subtract from the culture is afforded much more protection than borrowing to build on it. Only the parodist can write or create with confidence. An artist or a writer can borrow elements from a copyrighted work if she is directly criticizing the original work. If the second work simply uses the original work in a comedic or satirical fashion for the purposes of spoofing something else, such as the mores of society at large, fair use would not protect the new artist.[4]

The argument that *The Wind Done Gone* is a parody of *Gone with the Wind* did not sway the district court, which issued an injunction against publication of the novel, which Houghton Mifflin immediately appealed. The federal appeals court promptly vacated the injunction, declaring from the bench that the injunction was a clear breach of the First Amendment of the U.S. Constitution, which protects freedom of speech, religion, and press. The novel made it to bookstores in weeks and sold briskly. The two parties eventually settled the case out of court.

The story seemed to end well: Americans read a temporarily banned novel that spoke to their history and social fabric from a much needed perspective, and Houghton Mifflin was free to recoup its investment in the book. Margaret Mitchell's estate retains control over any nonparodic representations of Rhett Butler and Scarlet O'Hara.

It's not clear who is the anarchist and who the oligarch in this legal conflict between a big publishing house and a major bank. However, we can study this case to examine how culture works, and how the fallout could injure the very habits and practices that built important works. Culture is anarchistic if it is alive at all. It grows up from common, everyday interactions among humans who share a condition or a set of common symbols and experiences. The collection of end products of culture—the symphonies and operas, novels and poems that survive the rigorous peer review of markets and critics—are often taken as the culture itself. Instead, culture is the process that generates those products. If it is working properly, culture is radically democratic, vibrant, malleable, surprising, and fun. These two visions of culture explain the different assumptions behind information anarchy and information oligarchy. Anarchists believe culture should flow with minimal impediments. Oligarchs, even if they seem politically liberal, favor a top-down approach to culture with massive intervention from powerful institutions such as the state, corporations, universities, or museums. These institutions may be used to construct and preserve free flows of culture and information, but all too often they are harnessed to the oligarchic cause, making winners into bigger winners and thus rigging the cultural market.

Overregulation risks cultural stasis. Winners can silence losers and late starters, which almost happened in the *Wind Done Gone* case. The estate of Margaret Mitchell, thanks to *Gone with the Wind*, is a cultural winner and has been since the book's debut in 1936. Alice Randall was an unknown writer before the publication of *The Wind Done Gone*. Had Randall's novel never seen a commercial bookstore shelf, it surely would have found a way through underground channels. And while underground channels would have allowed the connected—those with the curiosity, motivation, and cultural capital to seek out and share the text—they would not have pushed the work into public view, where it could have some measurable effect. *Gone with the Wind* would have remained the last official word on the Tara saga. And messing with the myth would have been considered as forbidden as trespassing on a plantation.

Everything Can't Be a Parody

Although the story ended well for Randall and her novel, it does not end well for the culture in general. Randall's victory is an anomaly; she was fortunate enough to have the legal and commercial power of a major publisher on her side. The ugly truth is that the courts should have stopped Houghton Mifflin from publishing *The Wind Done Gone*. If we take the law seriously, we must grant the Mitchell estate complete control over all subsequent creative uses of the plot and characters for the duration of its copyright term. The law clearly reserves for the copyright holder rights over all "derivative works" that may emerge from the original. Defining parody narrowly and precisely, the district court read the new novel correctly. *The Wind Done Gone* is not a parody of *Gone with the Wind*. It is a revision, a retelling of the classic from a critical stance in a vernacular voice. It's also a supplement, another perspective in the parallax view of race and history that the United States desperately needs in order to comprehend and acknowledge its own tragic nature.

We deserve the ability to read Alice Randall's book. Yet the law, properly invoked, does not let us. The problem does not lie in the nature of Randall's work but in American copyright law: It is censorious. The law is unjust. Instead of cheating by stretching the definition of parody to include commentary and revision posing as parody, the appeals court should have considered the big questions: Does American copyright law protect works for too long, in too many ways, and with too many ancillary powers for the general good? Wouldn't American culture be better off if we were free to revise works that mean so much to us, even if we want to do it out of love and respect instead of derision and ridicule? The court should have explored the clash—clearer every day—between copyright's power to squelch speech and the First Amendment's duty to allow it. Instead, the appeals court, like many courts before it, evaded the tough questions. Had courts consistently supported the publication of an important novel from a major commercial publisher, then either the U.S. Supreme Court or the U.S. Congress would have had to confront the censorious nature of copyright and rethink some recent changes in the law, which have tilted power in the cultural markets in favor of established voices and against emerging ones.[5]

Rigging the System

By any reasonable measure, recent changes to expand the power and lengthen the term of copyrights in the United States have been a complete failure—doing nothing to stem real piracy and nothing to prevent widespread file sharing. Yet they have burdened scientists, librarians, scholars, students, and engineers. They have chilled some political speech, art, and Web linking. These radical changes have been hard on the good guys and irrelevant for the bad guys.

The Digital Millennium Copyright Act (DMCA) is misnamed. It's actually the "Digital Millennium Anticopyright Act." It replaces a human-based, democratically generated system with misplaced faith in cold, blunt technology. Where once users could assume wide latitude in their private, noncommercial uses, now a layer of code stands in the way of access to the work itself, preventing a variety of harmless uses. Because the DMCA allows content providers to regulate access, they can set all the terms of use. The de facto duration of protection under the DMCA is infinite. While copyright law in 2001 protects any work created for the life of the author plus seventy years (ninety-five years in the case of corporate "works for hire"), electronic gates do not expire. This allows producers to "recapture" works in the public domain. This also violates the constitutional mandate for copyright laws that protect "for limited times." The DMCA works over and above real copyright law.

Most dangerously, producers can exercise editorial control over the uses of their materials, extracting contractual promises not to parody or criticize the work in exchange for access; many Web sites already do this. Just as dangerously, the DMCA could enable producers to contractually and electronically bind users from reusing facts or ideas contained in the work. Despite the ineffectiveness and counterproductive effect of the law and the technology it supports, the copyright industries insist on defending the DMCA as if their future depended on it.[6]

From Software to Hardware

Despite its failure to protect music and video, the DMCA is more important than ever in your garage, office, and living room. Hardware industries (industries outside what we generally consider "software" or copyright industries like film, music, text, and computer code) are increasingly using the DMCA to lock in monopoly control over secondary goods. These goods have nothing to do with copyright, nothing to do with creativity, knowledge, or art. Because it's possible to put a computer chip in almost anything, companies are doing so. And when a company puts software on a chip that sits on a removable part of a machine, and puts some other software on a complementary chip in the larger device, the DMCA prevents another company from developing a replacement for that part.

Consider, for example, garage door openers, which rely on a remote control device. For decades these devices worked by emitting a high-frequency tone specific to the brand. Within range, a driver could open her garage door from a warm, dry car. If she lost the remote control device, she bought a replacement from the hardware store. Because these tones were open and easy to decode, several companies offered generic replacements for the openers, keeping the replacement price low. But since 1998, it has been possible (in fact advisable) for manufacturers to install a bit of computer code in their remote control devices and receivers so that only controllers with the right code can open the door. The code serves as a "handshake" between controller and opener, tethering one product to the other and thus limiting competition. When a company called Skylink Technologies introduced a generic digital remote controller in 2002, the garage door opener manufacturer Chamberlain Group sued, claiming a DMCA violation. In a similar case, Lexmark International, the maker of inexpensive printers for home computers, installed an electronic handshake between its printers and the toner replacement cartridges in an effort to lock out competition. Lexmark is suing Static Control Components for selling a replacement cartridge that circumvents this access control. In neither case does the access control device protect creative content. Yet the DMCA is written so broadly that it applies directly to such cases. It's not hard to imagine an automobile company installing handshake technology between its batteries and its battery cables or its dashboard and its car stereo system to prevent third-party competition and consumer customizability. Every time you needed a new battery, you would have to go to the original manufacturer—or risk committing a federal copyright violation. If your car manufacturer went out of business or stopped making a water pump for a particular vehicle, you would have to go to a junkyard or buy a new car.[7]

The law has reached an intolerable level of absurdity when participating in a pirate economy is easier than ever while participating in a legitimate, competitive economy is harder. This situation stems from a widespread anxiety that digitization and networking can wreak havoc on our industries. To some extent they have. To an equally important extent they have locked winners into place and stifled the free market. This is all part of a widespread corruption and entanglement of the bundle of laws we colloquially yet inappropriately call "intellectual property."

Intellectual Property Is Neither

Intellectual property is distinct from "real" property because it is not naturally scarce. If someone steals my car, I am left with no car. Yet if someone photocopies my book, I still have my book. The fundamental purpose of intellectual property law is to create artificial scarcity.[8]

The intellectual property concept is quite recent. Some accounts trace the origin of the phrase to the 1930s, others only to the 1960s. The phrase did not fully enter public discourse until the 1970s, when it was introduced to sum up a variety of efforts to make sense of changes in the global flows of culture and technology. Note that the various branches of law included under the umbrella of intellectual property have distinct purposes, subjects, and histories, some centuries old. Patents generally govern the use of the ideas behind inventions and useful processes. Copyrights deal with creative goods in various media and thus have profound implications for free speech and democracy. Trademarks protect a company's name and reputation in the marketplace, signifying consistency (if not quality) in the good being marked and marketed. Laws concerning trade secrets emerged from a concern to protect companies from espionage or extortion by former employees. Right-to-publicity laws gained importance as celebrity images became more valuable in the latter half of the twentieth century. European governments have recently endorsed a new form of intellectual property law to fill in a perceived gap that copyright does not fill: protection of collections of facts and data.

These areas of law grant limited monopolies to creators, inventors, discoverers, or marketers to exploit their goods with reduced competition. In each case, the law offers some incentive for creators to invest in their products or processes. The assumption that justifies the laws is that without such state-granted limited monopolies creators would have little incentive to produce and distribute. Others would soon run away with the fruits of their work and creativity, causing the price the products could fetch to fall precipitously. Balanced against the incentive to create is the idea that the public should have an opportunity to exploit the works and products after the monopoly has worked for some period of time. At least in the case of patents and copyrights, which have statutory limits on their duration, works eventually enter the public domain and become much cheaper and easier to build on for future products. Because of the public domain, schoolchildren can read the works of William Shakespeare in inexpensive editions and low-income people can purchase generic versions of once expensive pharmaceuticals.

The major elements of intellectual property—copyrights and patents—emerged in Europe before the industrial revolution. Some argue that they enabled the industrial revolution. By the mid-nineteenth century, most European nations and the United States and Canada enjoyed domestic copyright and patent protection but did not respect each other's protections. Pirates in one country exploited the works of authors and inventors who were protected in their home countries. Charles Dickens complained about cheap pirated American editions of his works. Mark Twain railed against cheap British and Canadian editions of his books. By the end of the nineteenth century, most industrializing nations (not the United States, which was the greatest of pirate nations until the 1980s) had agreed to respect each other's copyright and patent regimes. By the end of the twentieth century, the United States and Europe sought to standardize intellectual property across the globe as more nations achieved literacy and industrialization. American and European companies seeking new markets did not want their products copied in countries lacking intellectual property protections. So the developed world pushed for the establishment of the World Intellectual Property Organization (WIPO) and the Trade-Related Aspects of Intellectual Property Rights (TRIPS) accord. WIPO members generate treaties and agreements about global intellectual property standards. Signatories of the TRIPS accord may, through World Trade Organization enforcement mechanisms, seek retribution for violation of intellectual property standards.[9]

Globalization and standardization efforts have generated consternation among developing nations, where farmers resist limits on the use and replantation of patented seeds and plants, and embryonic publishing and media companies are hard-pressed to succeed when they must play by rules written by their more powerful and established global competitors. Northern concerns that developing nations serve as havens for software and video pirates have kept pressure on their governments to adopt and enforce laws that resemble those of the United States and western Europe.[10]

While the rise of digital technologies and global communication networks has generated a crisis of sorts for intellectual property owners, it has also generated a significant opportunity to expand control over products and uses far beyond the limits allowed by traditional law. When the representatives of developed nations considered the potential for widespread global piracy of copyrighted materials and Internet pranksterism in the shape of bogus World Wide Web sites, they created mechanisms to limit such activities. Several countries, including the United States, have passed laws forbidding the distribution of any technologies—even simple mathematical algorithms—that might evade or crack access or copy control mechanisms. Digital rights management technologies protect not only copyrighted material but also material that is already in the public domain and facts and data that are not covered by copyright law. Digital lockdown grants far greater control over works than traditional copyright law ever did. Copyright enforcement is leaving the realm of human judgment and entering a technocratic regime. Through quasi-governmental regulatory institutions such as the Internet Consortium of Assigned Names and Numbers (ICANN), those who control trademarks or publicity rights enjoy the power to shut down Web sites that parody or criticize them, and famous people may even squash sites run by nonfamous people who share their names.

The Copyright "Crisis"

Conventional wisdom posits a crisis in the enforcement of intellectual property in a wired world when actually there is a crisis of confidence in intellectual property. It's a cultural crisis and a crisis about culture, which are not the same. This crisis of confidence, more than the technological or legal challenges of the past twenty years, is what threatens the remarkably successful regulatory regimes of copyright, patent, and trademark law. These areas of regulation have become the central battleground between the forces of anarchy and the forces of oligarchy. Systems that were until recently reasonable, democratic, and beneficial now are, more than contested, untenable.

Advocates of maximum global protection have undermined public confidence in the premises and practices of these systems. They have exploited the anxieties of the digital moment and the global age and have parlayed their growing economic power within industrialized nations into raw political power. But those who want to use cultural products in traditional ways and those who express concern about values not embodied by the maximalist position find themselves alienated. So people by the millions are flaunting their ability to share culture and information, and they are steadily losing faith in the ethical foundations of copyright. In trying to exert an absurd level of control over culture and information, the intellectual property industries undermine their own cause. While installing cumbersome locks and gates, they have undermined the norms on which adherence to the law depends.[11]

Jack Valenti, chairman of the Motion Picture Association of America until he resigned in early 2004, is among those who must accept blame for eroding real copyright law by undermining public support for it and placing his hopes in technology. He refuses to be satisfied with the radical changes that shifted the balance of copyright away from consumers and toward copyright holders. Now he wants the U.S. Congress to reengineer the information infrastructure once again. In May 2002 congressional testimony Valenti said, "The facts are these: the copyright industries are responsible for some five percent of the GDP of the United States. They gather in more international revenues than automobiles and auto parts, more than aircraft, more than agriculture. They are creating new jobs at three times the rate of the rest of the economy. Brooding over the global reach of the American movie and its persistent success in attracting consumers of every creed, culture, and country is thievery: the theft of our movies in both analog and digital formats."

Valenti blamed hackers, universities, and China for the woes of the American film industry—woes that seem hard to credit in light of his opening statements about his industry's global success. Valenti was demanding government protection against the perils of the free market and the global information flow. Specifically, Valenti asked Congress to consider two bills that would severely alter the information ecosystem: the Consumer Broadband and Digital Television Promotion Act (CBDTPA), which would require hardware manufacturers to prevent unauthorized digital reproduction, and an as yet unarticulated move to embed watermarks on analog signals that emanate from digital content and force hardware manufacturers to prevent unauthorized recording of these analog signals. Valenti wants to use the power of the U.S. government to restrict the machines that Americans can make, sell, and use, all to protect his little industry.[12]

These are the actions of a Constanzan cynic. Valenti makes narrow, selfish arguments on behalf of the copyright system. He does not acknowledge that copyright enriches the public sphere or enlightens the republic. To Valenti, copyright is all about return on investment, and therefore he wants to shift it from a protocol-based system to a control-based one.

Some claim Valenti's view of copyright is a relic of the industrial age, when expensive printing presses took a substantial investment to purchase and maintain. Only a few could afford to produce and distribute cultural material, so the state had to create an incentive system—a limited monopoly granted by copyright. Richard Stallman founded the Free Software Foundation and is a leading cynic (in a Diogenic sense) in regard to copyright. He argues that "copyright in the age of the computer network is a restriction on the general public, not just an industrial regulation. The economies of scale are vanishing—anybody who can use the material can copy it about as well as anybody else. There is no reason why copies have to be made centrally and shipped in physical form. It's not necessary or useful anymore, except to those who want to funnel all the copying through their tollgate. As a result, copyright's role has been completely reversed. It was set up to let authors restrict publishers for the sake of the general public. Digital technology has transformed it into a system to let publishers restrict the public in the name of the authors."[13]

In this passage, Stallman speaks globally, not just as an American, and those concerned with the greater good might agree with him. Richard Stallman is a true cynic. As a person who strictly lives according to his beliefs, a man who values freedom over convenience, commerce, and control, Stallman lives like Diogenes. More importantly, he lives according to the principles expressed by William James and Francis of Assisi, both of whom incorporated aspects of the Cynical life into their own. Like Diogenes, Stallman is sometimes disagreeable, ill-mannered, and blunt. Also like Diogenes, he is deeply concerned with the health of the global polis, and the freedom of the individuals in it. He revels in exposing hypocrisy wherever he finds it.

The only problem is that he's wrong. Real copyright should still be relevant in the digital world. Stallman fails to consider the values that might be enhanced by a system like copyright even in an age when the cost of distribution of digital goods approaches zero. Reducing the cost of copying and distribution does nothing to change the cost of production, editing, and publicizing materials. Although Stallman seems to ignore it, the fact is that much content still exists in analog form for reasons of stability and usability. While copyright has been hijacked by interests stronger and richer than individual creators, it can empower creators when they have the knowledge and fortitude to fight for their rights and bargain from strength. Stallman's vision of creativity involves collaboration among many authors, each willing to forgo upfront capital for a later payoff: social capital, cultural capital, or political capital. Stallman believes passionately in writing good code for the sake of making the world better and freer. Yet he has not made space in his model of creativity or his critique of copyright for less altruistic modes of creation and distribution, like this book.

The Internet, since its inception, has been more hospitable to Richard Stallman than to Jack Valenti. Yet Valenti's vision of information regulation has triumphed at every level—global, national, and personal. So let's look at what Jack Valenti has wrought. Let's look at what has happened to copyright in the past decade, as digital technologies and networking have radically changed the terms of discussion. The domain of copyright regulation has shifted from the humane and democratic to the technocratic and dictatorial. As Richard Stallman points out, the terms of access, distribution, and copying now are dictated by the copyright holder, not negotiated through the market or the law.

Simultaneously, and not coincidentally, there has been a massive copyright revolt. Every time a major media company rolls out a new digital protection scheme, within a month someone publishes a hack of it. In the late 1990s, Jack Valenti and his cohorts in the various content industries abandoned copyright. They decided that a porous, imperfect regulatory structure was no longer relevant or optimal for their businesses. So with minimal public input and reflection, they convinced the World Intellectual Property Organization (WIPO) and the U.S. Congress to scrap real copyright in favor of technological regulation. Yet they still summon the gumption to defend the ethics of copyright adherence. They want to have it both ways. They want to abandon the democratic safeguards of copyright such as fair use and a rich public domain while the public romantically embraces the prohibitions embodied in copyright.

Valenti's vision is oligopolistic. He wants a handful of companies and states to dictate the terms of exchange for digital content. Stallman's vision is anarchistic. He wants freedom—convention, stability, and predictability be damned. The nature of the Net favors Stallman. The nature of state and corporate power favors Valenti. A tamer, more reasonable, more manageable vision of our information ecosystem seems harder to grasp every day. I hope the future of copyright lies somewhere between these extremes. In a perverse way, Valenti's technocratic abandonment of real copyright—cynical in the George Costanza sense—undermines the social contract that supports the copyright system. Until we come up with a set of reasonable protocols—an ethical cynicism in the Diogenic sense—we will find ourselves losing all the benefits of a reasonable copyright system.

Copyright as an Instrument of Censorship

The purpose of American copyright is expressed in Article I of the U.S. Constitution: "to promote the progress of science and useful arts." Contrary to popular belief (and European principles) copyright is not primarily meant to grant control and benefits to authors and artists. Historically it's a publisher's law more than an author's, and the ultimate beneficiary is supposed to be the public. American copyright emanates from this clause, which directs Congress to create a federal law that provides an incentive to create and distribute new works. The law grants an exclusive right to copy, sell, and perform a work of original authorship that has been fixed in a tangible medium. The monopoly lasts for a limited time and is restricted by several provisions that allow for good-faith use by private citizens, journalists, students, and scholars. Copyright was created as a policy that balanced the interests of authors, publishers, and readers. Copyright is a "deal" that the American people, through its Congress, made with the writers and publishers of books. Authors and publishers would get a limited monopoly for a short period of time, and the public would get access to those protected works and free use of the facts, data, and ideas within them.[14]

But over the past decade, Congress has allowed content providers to dictate radical changes that have undermined these democratic safeguards. As it stands today, American copyright stifles real creativity by everyday people. It no longer promotes progress. It protects the established and handcuffs the emerging. American copyright, which filled our libraries and served as a brilliant engine of creativity, has been corrupted to the point of counterproductivity.

Contrary to Stallman's protests, real copyright is not outdated. We still need it. If there were no copyright laws, unscrupulous publishers would simply copy popular works and sell them cheaply, cutting out the author and legitimate publisher. Creators and distributors still depend on the stability the monopoly power of copyright grants them to justify investments. But just as importantly, the constitutional framers and later jurists concluded that creativity depends on the use, criticism, supplementation, and consideration of previous works. Therefore, they argued, authors should enjoy this monopoly just long enough to provide an incentive to create more, but the work should live afterward in the "public domain," as common property of the reading public. A monopoly price on books was considered a "tax" on the public. It was in the best interest of the early republic to limit this tax to the amount that would be sufficient to provide an incentive, but no more and for no longer than that. This principle of copyright as an incentive to create has been challenged in recent decades by the idea of copyright as a "property right." Therefore, many recent statutes, treaties, and copyright cases have seemed to favor the interests of established authors and producers over those of readers, researchers, and future creators. These recent trends run counter to the original purpose of American copyright.[15]

When copyright worked well, it was fueled by widespread respect for the dignity of creators, a healthy respect for and understanding of fair use, and a general sense of fairness about the players in the systems of production and distribution. When technologies failed to allow all but a few into markets of distribution, and few of us had the power to make copies for ourselves, these norms were self-evident. They did not need grooming, revision, or rearticulation. The system was not perfect, just workable. We need to restore these norms or generate new ones that account for the remarkable democratic power many people now have over their infoscapes and mediascapes.

Taking the spirit of that constitutional clause seriously, we should construct a regulatory system that grants creators and publishers a limited monopoly that provides just enough incentive to produce and distribute materials. If we demand too high a "tax" from readers and the next generation of creators, then we risk creating a counterproductive and restrictive regulatory system, one that retards science and art. And that's exactly what we have done. The copyright system has been extended, twisted, supplemented, and corrupted to such an extent that the oligarchs have locked up content in most forms for what seems like ever. Law-abiding, reasonable users find themselves devoid of rights or abilities to reuse or build on what came before. Only the anarchists—those who directly challenge the foundations of copyright—have strengthened the ability to acquire and use elements of culture. In desperation, more and more of us are embracing information anarchy. But this only leads to a Hobbesian black market of culture and information, one we can already glimpse by peering into economies and societies undergoing massive change.

CHAPTER 7

Culture as Anarchy

Most of what we think we understand about peer-to-peer communication and battles over the control of culture and information rests on the American experience. Most of the debate concerns the competing American values of free speech and growing commerce. Soon (if not already) nation-states like India, China, and Mexico will play a larger part in the global dynamics of trade, creativity, and free speech. In these areas, where the vast majority of the humans born in the twenty-first century will reside, we find the starkest contrast between information anarchy and information oligarchy. Not coincidentally, we also see the human detritus of the neoliberal economic project, which undermines indigenous industries, social safety nets, and traditional cultural practices while latching billions to the fickle winds of global trade. The rhetoric surrounding these issues is troublesome enough in parts of the world with a rich tradition of respect for copyright law and plenty of disposable cash for legitimate products. But it assumes an absurd tone in places where the vast majority of middle-class people cannot afford the cultural products they need to be citizens of global culture or players in the global economy. If soft, humane methods of information regulation fail in the "developing" world, there is not much hope for them.

In New Delhi, India, the pirate industries have inherited some well-known marketplaces around the crowded, cosmopolitan capital city. During most of the past decade, customers went to Nehru Place to find pirated copies of Microsoft Windows or Adobe Photoshop. There are even reports of pirated copies of Red Hat Linux operating system discs, even though Linux is ostensibly "free software."[1]

Indian professionals can't afford to buy legitimate software. A design consultant in Delhi might earn 100,000 to 600,000 rupees per year. Legitimate

copies of the design software she needs to do her job (Photoshop, Free-
hand, PageMaker, etc.) would cost her about 400,000 rupees. Pirated versions cost about 1,000 rupees in Nehru Place. Businesspeople are not the
only consumers of software. Indian education might have to give up
dreams of computer literacy for its students if it had to pay market prices
for software.[2]

Everybody in Delhi knows about Nehru Place. It's in all the papers and
on everybody's lips. It thrives thanks to corrupt police officers and a general
acceptance of the black market economy. Occasionally the municipal police
stage a raid for show, especially if state or national leaders want to reassure
American diplomats or business leaders that they are dealing with
the piracy problem. But most of the time there is scant copyright enforcement in bazaars like Nehru Place or Palika Bazaar, the place to buy pirated
CDs, VCDs, and DVDs. In the 1980s, Palika Bazaar was the center of
pirated audiocassette culture and commerce. Soon digital entrepreneurs
replaced the analog vendors. On most days in these places, public enforcement is either invisible or irrelevant. The police are paid well by the pirates.
Increasingly, however, the software and entertainment industries are hiring
private enforcers, often retired police officers, to deal with the vendors.
Increasingly violent private-sector raids on Nehru Place and Palika Bazaar
are causing the vendors to redistribute themselves around the city. There
has been a shift from single commodities in centralized marketplaces to
multiple small outlets selling a wider variety of pirated digital goods. There
are cottage industries in every area in every city in India. Madipur Colony
on the outskirts of Delhi has become one of biggest centers of CD writing.
Every house seems to have a CD burner in it. Customers move from house
to house collecting the titles they want. There is no way to shut this down
short of arresting or beating every family. Madipur Colony is the physical
and geographical embodiment of Gnutella, a distributed market that defies
the surveillance powers of the state and corporate raiders. The terminology
of state enforcement and the global market can't make sense of this increasingly anarchistic process. When laws don't matter, oligarchs have teams of
thugs armed with billy clubs.[3]

India has had copyright statutes since 1957, but media industries actively demanded enforcement only since about 1992, with the advent of
globalization, digitization, and liberalization. As Bangalore lawyer
Lawrence Liang observes, "The need for enforcement has translated into a

public–private partnership between the info–entertainment industry and the state."[4]

The Indian government has incentive to crack down on piracy and pursue video pirates abroad. The Indian film industry, popularly known as "Bollywood," has been hemorrhaging now that its audience is growing far beyond the South Asian subcontinent and its substantial diaspora. Indian films are the rage all over the world, and the most popular type of film (and the model of an emerging popular culture industry) in Nigeria. Let's consider my film format choices if I lived in northern Nigeria, which is unlikely to get a Blockbuster video store anytime soon. Lacking access to the Internet, I could not use Amazon.com. I could only get this film at my local pirate video shop. These establishments, which are sprouting up along ancient trade routes in North Africa, Southeast Asia, and the Indian subcontinent, are an important element of everyday life in much of the world. The proliferation of piracy bazaars and shops that sell unauthorized copies of Indian films has left Bollywood frustrated. Revenues were down more than $30 million in 2002, compared to 2001. While some blamed a worldwide economic slump and a dearth of hits, others blamed piracy. Regardless of the source of losses, Bollywood (as well as Hollywood) is alarmed about the normalized distribution and consumption of pirated material in much of the world.[5]

As anthropologist Brian Larkin reports, northern Nigeria is a pirate's paradise, and Indian films are wildly popular. The piracy infrastructure in Nigeria, Larkin explains, exemplifies the entrenched place of piracy in many parts of the world. Nigerian domestic producers also claim to be suffering as a result of the proliferation and popularity of pirated Indian and American media products in the past decade. Nigeria underwent a massive transformation during the 1980s and 1990s. Military dictators rose and fell, and a fragile democracy finally emerged. The Islamic northern section of Nigeria demanded laws that differed from those in the Christian south. Ethnic and religious tensions boiled over into violence. Much of this change was generated by the changing role Nigeria plays in the global economy, and the changing role the rest of the world plays in Nigeria. The former British colony has become a leading petroleum exporter; it has also become a node in global smuggling operations. Massive numbers of Nigerians have emigrated to Europe, Asia, and North America. Satellite television and mobile telephones allow Nigerians to stay in contact with their friends and family members in Houston, London, and Los Angeles. Money and cultural products flow throughout these networks.

However, the signal is distorted. The quality of voice and image, music and video, is quite poor. Most videos are copies of copies of copies of poor original versions. Machine breakdown is as important in Nigeria as the introduction of technology itself. Because the Nigerian mediascape constantly needs repair, the hacker, the tinkerer, is increasingly important. Those who repair equipment often master the skills of dubbing, editing, remixing, and distributing video from multiple sources. Everything is distorted. Everything is cheap. Everything is illicit. And everything is either breaking or being fixed. Piracy and hacking are the source of media experiences for much of Nigeria. There is no other distribution infrastructure. A more expensive form of content or mode of dissemination would not succeed in a environment in which everything breaks down, especially the electricity supply. As Larkin points out, piracy enables media flows outside the view of the state, which is often hostile to the content or producers. The pirate network is so diffuse and ingrained that there seemingly is no way to stop it.[6]

In Mexico, a strong federal state is making the most of its partnerships with American companies. Still, legitimate media outlets may not survive in the hypercompetitive environment generated by blossoming pirate markets. The pirates may stamp out the record stores before the state stamps out piracy. In 2001 Mexican authorities conducted 302 raids that yielded 17.1 million illegally copied compact discs. Most Mexican pirated discs cost the new peso equivalent of $1. Legitimate discs cost more than $12. Mexico has more than 12,000 street vendors selling pirated compact discs, but fewer than 900 stores, which sell 57 million discs per year. As late as 1999, pirated discs accounted for about 15 percent of sales in Mexico. By 2002 authorities estimated that 61 percent of discs sold in Mexico were illegitimate.[7]

The Mexican piracy threat is very different from the silly U.S. panic over peer-to-peer file sharing, fair use, home recording, and archiving. Mexican pirates don't need high-speed Internet connections and sophisticated software. Peer-to-peer networks are useless to serious music pirates, whose needs are simple: a legitimate compact disc from a music store in Mexico or the United States, a $1,000 personal computer, a $30 music program, and a $70 compact disc burner. Like other growing nations with more than 100 million people, most Mexicans earn less than $5 per day. Yet they want to dance. Many want to listen to the same music and watch the same movies that their wealthier neighbors enjoy. People can't afford to feed themselves and buy legitimate compact discs.

The culture and information battles that matter to most people in the world are occurring on the streets and in the bazaars, not in the courts. Nothing that happens in courts or in computers is going to change that anytime soon. And current law enforcement efforts in disparate corners of the north are futile. There is no single gang that authorities can imprison to stifle global piracy. There is no single technology that governments can outlaw or companies can reengineer. There is no way media companies could price their goods to compete with black markets and informal networks when much of the world can barely afford the artificially supported American wheat that is being dumped on markets everywhere. What will it take to stop global piracy and restore price stability to the world's media companies? Nothing short of global economic stabilization and equalization, which means that Nigeria would become much wealthier and the United States much poorer. There is little chance that the United States government would let that happen, even if Jack Valenti asked it to.

Despite the obvious futility of antipiracy efforts, governments throughout the world are shifting to the technological regulation of distribution from the human to the technological, and expanding the jurisdiction from national to global. The responsibility for enforcement is moving from public to private. All of this is undermining what little faith the world had in the rule of law. The recording, film, and software industries gave up on real copyright in the mid-1990s when they decided that technology and private "self-help" were more useful to them than courts and social norms. Now they are forcing consumers to abandon humane regulation even as they plead for consumers to respect it. Since the mid-1990s these companies have pushed for global standardization of enforcement and protection mechanisms through laws like the DMCA and the proliferation of digital rights management schemes, regardless of the cultural and intellectual costs to global society. Give the spectacular failure of these measures in the developed world, how can they be expected to make a difference in Nigeria, Mexico, or India?

What kind of policies should developing nations construct to deal with the cultural needs and demands of their citizens? What kind of policies should the United States, Japan, and western Europe construct to recognize the complexity of global culture? Because the oligarchs have set the terms of debate and the reactions to them have been decentralized and negative, no one seems to know how to begin the conversation. We should be willing to learn some lessons from historical attempts to control culture and communication.

Muting the Beat

Drums were illegal in much of the American South before the Civil War. Slaveholders were aware of "talking instruments" in the West African tradition and tried in every way possible to stifle communication across distances. Drums could signal insurrection. Drums could conjure collective memories of freedom. Slaveholders realized that to subjugate masses of people, they had to alienate them from their culture as much as possible. The slaves were stranded in a strange land, and the slaveholders tried to make the land seem stranger than it was. They strictly regulated slave culture and outlawed slave literacy. They had to commit social and cultural homicide to keep otherwise free people from rising up and taking charge of their own bodies. That the rhythms of Africa and the Caribbean still set the time for American culture speaks to the determination and courage of African American slaves. The slaveholders outlawed the tools, but they could not stop the beat.[8]

As oligarchic forces such as global entertainment conglomerates strive to restrict certain tools that they assume threaten their livelihood, they should consider that, throughout history, people have used and adapted technologies in surprising and resilient ways. Once in a while, a set of communicative technologies offers revolutionary potential. Peer-to-peer networks are part of a collection of technologies—including cassette audiotapes, videotapes, recordable compact discs, video discs, home computers, the Internet, and jet airplanes—that link diaspora communities and remake nations. They empower artists in new ways and connect communities of fans.

The battle to control these cultural flows says much about the anxieties and unsteadiness of the power structures that hoped to capitalize on cultural globalization. It also has much to teach us about the nature of culture itself.

Global Culture by the Download

A couple of years ago, a journalist friend of mine put me in contact with a consultant who works for the World Bank. He called to invite me to participate in a meeting in New York that would put cultural ministers from a handful of African countries—including Nigeria, Ghana, and South Africa—in touch with leaders from the American music industry. The goal was to brainstorm how African musicians might exploit digital music distribution systems to market and deliver their songs directly to diaspora communities. He had no way of knowing what I thought of this idea. I had yet to publish anything on the subject and my opinions were not widely known. So he was not quite prepared for my reaction. "Why do they need record companies?" I asked. "The artists can do it all themselves for less than $10,000."

He was stunned. Having a World Bank perspective on development, he assumed that the artists of the developing world needed and would welcome the giant helping hand of Bertelsmann or AOL Time Warner. So he responded with an appeal to technological expertise. The artists would need the major labels, he said, because the labels are working on incorporating digital rights management software into digital music files. Without watermarking or copy-protection features, the artists would be giving their music away. I explained to him that it was too late for all that. Digitization and networking had beaten the record companies to it. I didn't even touch the complications inherent in asking African musicians—often dissidents—to work with government culture ministers. I just pointed out that he had missed a technological moment. Having the best of intentions, he did not consider that certain technological changes had fostered a new ideological movement as well. And these trends might change the nature of global music and creativity.

One of the great unanswered questions is how file sharing and MP3 compression will affect the distribution of what music corporations call world music, tunes from non-English-speaking nations offering rhythms that seem fresh to Europeans and Americans who have grown up on the driving four-four beat of rock and roll. Rhymes and rhythms from all corners of the Earth are now available in malleable form at low cost to curious artists everywhere. Peer-to-peer has gone global. Of course, there are some big economic and technological hurdles to overcome before it can affect all cultural traditions equally. As the differences narrow, how will the availability of a vast and already stunningly diverse library of sounds change creativity and commerce? Won't all music be world music?

On any given day, on any file-sharing system, you can find obscure and rare items. I have downloaded speeches by Malcolm X, reggae remixes of hits by Biggie Smalls, various club dance mixes of Queen's "Bohemian Rhapsody," and long lost Richard Pryor comedy bits that were released on vinyl by a company long since defunct. Through nation-specific and general world music chat rooms on the now defunct Napster, I found Tamil film songs, Carnatic classical music, and pop stuff from Asian Dub Foundation, Ali Farka Toure, Orisha, and Youssou N'Dour. The most entertaining phenomena of the MP3 libraries on peer-to-peer systems is the availability of "mashes"—new compositions created by combining the rhythm tracks of one song and the vocal track of another. The best example of a popular mash is "Genie's Revenge," a combination of vocals by Christina Aguilera and a guitar riff by the Strokes.

Ethnomusicologists are just starting to consider this phenomenon. During the 1980s and 1990s, anthropologist Steven Feld raised some serious questions about the future of global cultural diversity as world music gained market share and generated interest among Western producers and labels. Feld published some of his thoughts in an article entitled "A Sweet Lullaby for World Music." The article traces the development of marketing efforts for this new genre, which included drum beats from Mali to the ambient sounds of lemurs in Madagascar. Feld expressed concern early on that the very term "world music" separated some forms of music from what academics and music industry figures call "music." Since the rise of the world music genre as a commercial factor, music scholarship has been asking, How has difference fared in the new gumbo? Recent world music scholarship, Feld wrote, has revealed the "uneven rewards, unsettling representations, and complexly entangled desires that lie underneath the commercial rhetoric of global connection, that is, the rhetoric of 'free' flow and 'greater' access."[9]

"Free flow" is a controversial buzzword in north–south communication policy debates. Starting in the 1970s through UNESCO forums, the United States argued that the world community should establish standards that encourage the free flow of information across borders, ostensibly to spread democracy and civil rights. Many corrupt, oppressive states—chiefly India under Indira Gandhi—argued that the doctrine of free flow was merely a cover for what we now call the neoliberal agenda: sweetening American corporate expansion by coating it with the thin glaze of Enlightenment principles. The free flow versus cultural imperialism argument has unfortunately limited our vision and stifled discussions about what we might do to encourage individual freedom and cultural flow while limiting the oppressive and exploitative aspects of the spread of American and European modes of cultural production and distribution. In some circles this debate has been supplemented by approaches that emphasize the complex uses to which all audiences put cultural elements.

Some scholars, Feld wrote, embrace the "cultural imperialism" model. In contrast to those who raise concerns about the spread of new loud noises, "celebratory" scholarship emphasized the use and reuse of American and European musical forms in emerging pop sounds flowing from the developing world. It also celebrated the new market success achieved by artists from the developing world. This scholarship emphasizes the fluidity of cultural identities and predicts an eventual equilibrium of the power differences in the world music industry. This school, which I subscribe to, downplays the influence of hegemony and emphasizes the potential creative and democratic power of sharing. Instead of "celebratory," I prefer the term "gumbophilic."

Feld, who belongs among scholars who emphasize what he calls "anxious narratives," sees little possibility for resisting the commodification of ethnicity and musical styles. For the anxious, "global" becomes "displaced," "emerging" becomes "exploited," "cultural conversations" become "white noise." To make his point that we should not ignore the effects of the cultural violence of primitivism, Feld writes that for all the talk of "free-flows, sharing, and choice . . . the reality [is] that visibility in product choice is directly related to sales volume, profitability, and stardom."[10]

Feld is reflecting the specialized anxieties of ethnomusicologists. He is less concerned with the effects on the actual music and how it works in the lives of musicians and fans:

In the end, no matter how inspiring the musical creation, no matter how affirming its participatory dimension, the existence and success of world music returns to one of globalization's basic economic clichés: the drive for more and more markets and market niches. . . . We see how production can proceed from the acquisition of a faraway cheap inspiration and labor. We see how exotic Euromorphs can be marketed through newly layered tropes, like green enviroprimitivism, or spiritual new age avant-garde romanticism. . . . In all, we see how world music participates in shaping a kind of consumer-friendly multiculturalism, one that follows the market logic of expansion and consolidation.[11]

Though I celebrate sharing, free flows, and gumbo, I acknowledge the gravity of Feld's concerns. Despite his anticorporate stance, Feld is culturally conservative. Celebrating diversity and creolization is not enough. We must take account of the gains and losses of cultural connectivity via corporate channels. But my question now is: How does the globalization of the peer-to-peer imagination change these issues? Perhaps the spread of peer-to-peer libraries, whether through volunteer electronic networks or traditional piracy, should allay the concerns of anxious critics. Anarchistic music distribution—so far—has been all about decorporatization and deregulation. Music corporations no longer control the flow, prices, or terms of access. Music distribution, with entry barriers that are lower than ever before, offers the potential of direct, communal marketing and creolization. Widespread distribution of digital tools—recordable optical media like CDs, personal computers, and editing software—can empower creators on the margins of the world economy as never before. With that in mind, we should acknowledge some key concepts about cultural globalization: It's happening, but it's rolling out in ways that alarm those who hoped to profit the most from it; the prices and profits of globalization are falling unevenly and unpredictably; culture is not zero-sum. Using something does not prevent someone else from using it and does not degrade its value. In fact, it might enhance it.

If you assume that culture is zero-sum, that it can be "stolen" or "sold," then you can't acknowledge that culture is anarchistic and emerges from common, everyday interactions among people and peoples. Perhaps you see cultures as once sealed, now violated. Perhaps you see culture, as nineteenth-century English critic Matthew Arnold did, as a stable set of ideas and expressions that define a society and demand a defense by elites. In 1867 Arnold published a treatise called *Culture and Anarchy*. The book presented an extended argument of the cultural implications in John Stuart Mill's book *On Liberty* (1859). Arnold took Mill to task for endorsing a low level of cultural regulation. Culture, to Arnold, was all the good stuff that authorities such as himself said was culture. And culture in the Arnoldian sense was preferable to—in fact an antidote to—anarchy.[12]

Samuel Huntington expresses this same oligarchic theory of culture in his dangerously simplistic yet influential book, *The Clash of Civilizations and the Remaking of World Order*. Huntington sees cultures as grounded on certain immutable foundations. Cultural transmission, fluidity, and hybridity, he argues, are "trivial" when compared to the deep, essential texts and beliefs of a culture. Huntington emphasizes the role of the Bible in what he calls "western civilization" and the role of the Analects of Confucius in what he calls "Confucian civilization." But he disregards how the people who live in these cultures actually use the texts and symbols around them. "The essence of Western Civilization is the Magna Carta, not the Magna Mac," Huntington writes. Still, most residents of the nations he labels "western" have no idea of the history or significance of the Magna Carta, yet no one can underestimate the cultural power of the Big Mac. Huntington argues against cultural globalization, against flows and exchanges of ideas and information. He looks at a dangerous, angry world and prescribes walls instead of networks.[13]

Huntington's preferred world might be quieter, as well as darker and dumber. Cultures change, grow, and revise themselves over time if they are allowed to. And cultural life is healthier when cultures are allowed to grow and revise themselves. During the European Dark Ages (fifth to twelfth centuries A.D.) a large portion of the world severed its cultural arteries and relied on internal and local signs and symbols. Europe was stuck in a crippling cultural stasis while the rest of the world, led by Persian and Arab traders, moved on. The Dark Ages in Europe were a time of mass illiteracy and not coincidental concentrations of power among local elites. As Tyler Cowen explains in his book *Creative Destruction*, cultural exchange generates cultural change. Exchange might make disparate cultures more like each other, but it also infuses each culture with new choices, ideas, and languages. Exchange does not dictate how cultural groups might use or abuse these new raw materials and tools. Every area of the world becomes more locally diverse as long as people are free to borrow pieces of cultural expressions and reuse them in interesting ways.[14]

This idea of culture as temporal, contingent, dynamic, and creolized best describes how culture actually works in people's lives. No one lives in Matthew Arnold's culture. Few would want to live in Huntington's. The fact is, most of us don't have a clue why the Magna Carta as a document is important to us. Many more can explain how Madonna is important. And she is important in different ways to different people at different times. Like the culture that rewards her and follows her, Madonna is temporal, contingent, and dynamic. As Lawrence Levine explains, "Culture is not a fixed condition but a process: the product of interaction between the past and the present. Its toughness and resiliency are determined not by a culture's ability to withstand change, which indeed may be a sign of stagnation not life, but by its ability to react creatively and responsively to the realities of a new situation."[15]

If we use an instrument of technology or law to dampen the vibrancy of culture, we risk cultural stasis. Deployed carelessly, such instruments can freeze in winners and chill losers (or those waiting to play). But states and corporations use technology, law, incentives, and money to rig the culture game. Culture should be allowed to grow and change freely. But it rarely is.

How Do We Regulate Culture?

Some years ago at a conference called Prospects for Culture in a World of Trade held at New York University, I was seated at a table with an official from Canada's Department of Foreign Affairs and International Trade. We soon (predictably) found ourselves discussing the differences between our two nations. The official, a longtime trade policy expert, tried to explain how the Canadian government defines "Canadianness" in promoting and protecting indigenous music, film, and television. I grew up less than a half-hour drive from Canada; I have visited hundreds of times. I loved SCTV long before it appeared on NBC. I even learned to count to ten in French by watching *Sesame Street* on a Toronto station. But after all these years, I still can't understand what is essentially Canadian about Neil Young, William Shatner, or *The Kids in the Hall*. Canadians call shell-covered chocolates Smarties, and Americans call them

M&Ms. Canadians spell labor with a *u* and pronounce it with much more feeling than we do. That's about as far as I get. Thanks to my American obtuseness, my conversation with the Canadian official soon turned from cultural protectionism to trade protectionism—or so I thought. I initially thought those were two distinct subjects. But my Canadian friend saw them as one and the same. Several times, he even employed the phrase "American cultural policy."

I was taken aback. I did not know how to discuss "American cultural policy." If cultural policy existed, I would have heard of it. I have lived in the United States my whole life, engaging and embracing American culture with a passion that often frightens my immigrant father. Through years of graduate training and a doctorate in American studies, I have turned my passions into something that resembles a career, and I'm supposed to be an expert on American culture. Yet through all that television watching, NPR listening, air-guitar playing, history reading, Walt Whitman quoting, Elvis defending, and otherwise pompous conversing about the nature of American culture, until that day I had never encountered the phrase "American cultural policy."

At first I followed my gut instinct. I told the Canadian official: No, the United States takes a laissez-faire approach to culture—just what he expected an American to say. We love to believe that we have succeeded through entrepreneurship, gumption, and loopy optimism. Rarely do we concede that the state plays any role in the construction of what we value, including "private" universities, the anarchistic Internet, and other bits and pieces of culture. The very idea that the state might help "pick winners" offends our libertarian and republican sensibilities. I've always been allergic to easy answers, so immediately I began to question my initial assumptions.

Maybe we do have cultural policy. We don't have a U.S. secretary of culture, but we have a secretary of commerce. Skeptics might argue that American culture *is* commerce, but even without getting into that debate, we can at least ask whether the commerce department executes something that might be called a cultural policy. Canadians sure think so. I'd be willing to bet that French, Iranians, Serbs, and Brazilians do too. Non-Americans I have observed think that American cultural policy encompasses the various ways that the U.S. government bullies, wrenches, tricks, intimidates, and otherwise persuades other nations to let—among other things—American-made films open in their theaters. Recent American administrations put great pressure on other nations to let American films debut on more than a small quota of foreign screens and to standardize copyright enforcement in all media in all nations. Furthermore, the pact negotiated in the final days of the Clinton administration—to grant permanent normal trade status to China—will expose more Chinese consumers than ever before to the likes of Viacom programming.[16]

President Bill Clinton succeeded in some areas—radically expanding copyright globally, for example. But he failed to prevent "culture" from receiving special treatment in world trade disputes. Cultural productions, for instance, still have some exemptions from the normal rules for trade in the North American Free Trade Agreement (NAFTA). The George W. Bush administration clearly wants to pursue policy goals similar to its predecessor's as it tries to extend the model of NAFTA to the entire Western Hemisphere. But it too is likely to face opposition from nations with less politically powerful culture industries. What precisely are U.S. cultural interests? During the Cold War, the answer was easy. Through its various propaganda organs, the U.S. government paid to spread American culture around the globe; cultural policies were small, inexpensive elements of a grand, sweeping policy meant to contain and roll back communist expansion.

These days, American cultural policies are still part of a global vision, but culture is no longer peripheral. Commerce in cultural products accounted for more than 7 percent of the U.S. gross domestic product in 1999. Copyright-intensive industries like film, television, and music exported goods worth $79.65 billion, more than any other sector of the economy, more than the chemical industry, the aircraft industry, or agriculture. In the post–Cold War world, the enemy is no longer communism or socialism; it is the stubborn persistence of cultural sovereignty that stands in the way of American corporate expansion.

Still, the flavor of the culture being sold does not seem to matter to the U.S. government as long as it is being sold—not lent, borrowed, copied, or shared. The infrastructure matters more than the content. Striking down cultural protectionism could help chauvinistic films like *Pearl Harbor,* but free trade efforts might allow greater sales in developing nations of CDs by Ali Farka Toure of Mali and Cornershop of the United Kingdom. An American company, AOL Time Warner, distributes CDs for both global artists. AOL Time Warner also does not seem to care what language, instruments, and ideas are encoded on its CDs, as long as the CDs move from store shelves. That, Americans tell their critics, is why any U.S. policy that helps ease worldwide restrictions on trade and benefits American business also benefits business from other nations. In today's global marketplace, the national identity of a corporation means less and less. In 2001 the Canadian Seagram's Spirits & Wine was taken over by a French telecommunications company, in a three-way merger that created Vivendi Universal, which now owns Seagram's music and film archives (Vivendi later announced that it will acquire the American publisher Houghton Mifflin Company for $1.7 billion in cash). Policies that promote American-made cultural products help Japanese Sony, which produces many of the devices people need to use our products. Sony now produces films and music by American artists as well.

What seems like a nationalistic cultural policy from the outside looks like a content-neutral global economic policy from the inside. Sorting out cultural policy from economic policy can get complicated even when a country has overt rules for protecting or spreading its culture. Look at the complicated relations between Canada and the United States. Like some other nations, Canada has an explicit cultural policy, enforced by a cultural minister. For decades, the Canadian government invoked various tariff and postage policies to support indigenous magazines and deter American periodicals from flooding Canadian newsstands. American publishers exported to Canada the magazines they sold in the United States. Because they weren't focused on Canadian concerns, they didn't directly compete with Canadian publications for advertising money. In the mid-1990s, however, when Time Warner issued a magazine called *Sports Illustrated Canada* (mostly a copy of the version sold in the United States, with a few added articles on Canadian sports and lots of ads from Canadian companies), the Canadian government fought back: It placed an 80 percent excise tax on any ad revenues that "split-run" magazines earned in Canada. Then, in 1997, the World Trade Organization ruled that Canadian policy violated the General Agreement on Tariffs and Trade. That shocked many Canadians, who had been led into the trade liberalization game with assurances that culture would never make it to the table. To Canadians, the issue was one of keeping the public sphere at least somewhat Canadian. It was about the editorial content of periodicals. To the United States, it was about ad revenue and moving copies of a generic product. At bottom, the issues of money and culture were intertwined.[17]

If a nation adopts protectionist cultural policies when it fears for the future of its culture, then Canadian (and, similarly, French) cultural policies make sense. Multinational and American media companies have powerful megaphones and seductive products. Some nations understandably fear that unrestrained commerce could drown out local expression. But considered from a different angle, the fear that is at the root of French and Canadian policies also lies behind a high-profile example of an American cultural policy.

During the Clinton administration, under the leadership of Chairman William Ferris, the National Endowment for the Humanities tried to establish ten regional humanities centers to catalog, preserve, and celebrate the local cultural heritage of American regions. For many years, Ferris ran the Center for the Study of Southern Culture at the University of Mississippi, which was successfully defining the southern in southern culture, even as the Hard Rock Cafe was metastasizing to Atlanta and the French Quarter of New Orleans. The center served as a model for the proposed regional humanities centers. The new chairman, Bruce Cole, seems not to care about the centers and has allowed Congress to slash funding for them. As the NEH Web site explained at the time, "While we often talk about 'one nation,' the United States actually consists of many regions. . . . In a nation and world continually drawn together by travel, telecommunications, and the global economy, it is easy to overlook the importance of local experiences to our heritage and daily lives." The minister of Canadian Heritage couldn't have said it better.

In many ways, the chairmen of the NEH and the National Endowment for the Arts are the closest thing we have to secretaries of culture. Certainly both endowments are instruments of American cultural policy. They have protectionist goals. They finance and support American forms of expression that the market and the media often marginalize. Many, perhaps most, of their projects reflect an American state of mind, even if it exists in the distinct minority of American minds. The endowments may be woefully small, and they may have only nominal influence on the loud yawp of American culture. Like the Endangered Species Act, they may generate sensational headlines when they cross the wrong folks. Yet many of us would not want to live in a country without them. There seems to be an air of resignation about the fight to preserve regional heritage. Is southern culture (or, for that matter, Great Lakes culture) so static, so dead, so unappealing that state intervention is required to preserve it? Are we ready to stuff and mount local culture in a museum? We risk doing so if preservation efforts prevent it from evolving, recharging, and recapturing the public imagination. Above all, have we asked ourselves those questions about what is, in effect, a cultural policy? There are other examples of federal cultural policy as well, for example, the work of the National Park Service, the Smithsonian Institution, and the Library of Congress. Each agency chooses what to protect and what to discard.

Copyright law may be the most powerful instrument of global American cultural policy. It has altered local cultures, both domestically and internationally. The terms of its copyright system should reflect the form and function of all aspects of a country's expressive culture. Weak copyright laws make it almost impossible for creators to reap a profit from marketing cultural expression; copyright laws that are too strong (as I believe they are now) choke off creativity and scholarship. From the late eighteenth century through the beginning of the Clinton years, copyright provided for a very limited monopoly over expression: It was aimed at taxing readers just enough to provide a financial incentive to creators. Clinton's copyright policies, carried out through powerful international treaties, stacked the power of the federal government firmly behind large, established producers at the expense of users, emerging artists, and independent production companies. At the same time the Clinton administration was busy bolstering some copyright monopolies, it was simultaneously attacking Microsoft for unfair trade policies and carefully scrutinizing the merger that created AOL Time Warner. When the federal government scrutinizes one media merger but benefits another large media company, it is promoting winners and losers. It is executing an often unarticulated—perhaps accidental—cultural policy. While we may not have an official or overt American cultural policy, we certainly have American cultural policies. Under Clinton they were ad hoc, disjointed, uncoordinated, and contradictory. In many ways, that was for the best. If the federal government is going to mess with global and local culture, by all means, let it work at cross-purposes with itself. If American cultural policy were coherent, I fear, it would cohere to the Disney agenda—which few Americans and fewer Brazilians would support.[18]

If there is a theme to American cultural policy in the era of George W. Bush, it is to favor the big over the small, the corporation over the cottage, the global over the local. Bush's Federal Communication Commission tried to relax ownership requirements for broadcasters. The NEH moved from defending local or regional culture to investing in "great works." In culture, everything interesting and forward-looking happens on the margins. Our policies have severed the margins from the mainstream while anarchistic technologies have connected them in new and powerful ways. Envision a global cultural ecosystem composed of Blockbuster video stores and Madipur Colonies while public libraries wither from lack of support and interest.

CHAPTER 8

The Perfect Library

The forces that seek to capitalize on the fears engendered by openness, fairness, due process, and freedom have powerful weapons and formidable resources at their disposal. They especially like to limit discussion by declaring an emergency. So the Enlightenment is always on the defensive, always saying, "Slow down. Think this through. Talk this out. Don't rush to judgment." This is not a powerful political rallying cry. It doesn't stir the soul or bring tears to anyone's eyes, or fill our streets with cries of protest. It doesn't make anyone rich or powerful.

But the Enlightenment can shine through in the most surprising places. On one of the last evenings of 2001 I sat in a dark theater on Gezira, a small island on the Nile in the center of Cairo, to watch an Arabic translation of Bertolt Brecht's tragic play *The Life of Galileo*. The authoritarian government of Egypt had no reason to allow the production of this play. It's about the power of truth and the depravity of censorship and torture. It's the story of one of the heroes of the European Enlightenment. While I watched the play (understanding no dialogue but the word *kitab*—"book" in both Arabic and Hindi), I considered how the Egyptian government had frightened its dissident thinkers—from Muslim fundamentalist clerics to liberal democrats—into exile or submission through torture and arbitrary prosecution. I thought about the occasional roundups of homosexuals. I pondered how the Egyptian government deflects popular discontent onto other targets like Israel and the United States. It also occurred to me that Egypt is one of the few authoritarian one-party states that shows no interest in censoring the Internet. I thought I would feel some sort of dissident thrill attending such a liberal play in such an oppressive atmosphere, but I didn't. There were no security forces monitoring the performance. The crowd seemed relaxed and slightly bored (many others did not grasp the dialogue fully because it was in classical Arabic, not Egyptian). The more I thought about it, the more I realized that this government had no reason to fear the message of Brecht's Galileo. The Egyptian government is aggressively secular and its greatest threats are from religious zealots. Despite Galileo's need to express himself freely and speak the truth, the play basically tells the story of Galileo's failure. In the end, the great scientist breaks under the pressure of his theocratic abusers. If anything, the play teaches us that resistance is futile, and that those who torture in God's name are the greatest villains.[1]

Leaving the island and walking through central Cairo to the Garden District with all the foreign embassies, I passed a dozen armed guards posted on street corners. I was reminded once again that I was visiting an oppressive, defensive state. The rarified air of the theater means nothing to those battling for control of Egypt's future. The stakes in Egypt are higher than the desire to control words in classical Arabic voiced on a small stage. The streets are in play. The theater is for those wealthy and comfortable enough to be distracted from the real battles over Egypt's future.

Cairo resembles like-size cities around the world more than it does the rest of Egypt. The dramas that play themselves out daily on its streets presage conflicts likely to spread to every corner of the Earth: teeming crowds of disgruntled citizens watched by armed guards. The press is remarkably free and mildly criticizes the government occasionally, yet illegal religious sermons by outlawed clerics pass from hand to hand on home-copied cassettes and compact discs. The mosques overflow on Fridays. Kentucky Fried Chicken overflows every day.[2]

Cairo is cosmopolitan for those who can afford cosmopolitanism. It's provincial and stratified for those who can't. It's one of the hubs in the ever widening network of global labor flows, with new residents from all over the Middle East and Africa trying to make lives in the slums and garbage piles that skirt the edges of the expanding city. It's also the high-tech and economic center of the Arab world. Its major faith preaches brotherhood and peace yet persists in discriminating against its darker-skinned or Christian minorities. Women may wear skirts and pants, yet many choose to cover themselves completely. The streets meant for its residents are filthy. The streets meant for tourists are clean. There are few animals running free in the city, but the air is thick with pollution. Many foreigners come to Cairo to imagine the ancient past in the shadows of the pyramids or the glass and gold of the King Tutankhamun exhibit at the Egyptian Museum. I could not help thinking that I was glimpsing the future. The rest of the world is becoming a little more like Cairo every day. Cairo, the hometown of terrorist Mohammed Atta, breeds anarchy and feeds oligarchy.

Three hours north of Cairo by train sits Alexandria, a symbol of all that could have been. Alexandria was twice (and might again be) a center of cosmopolitan, multicultural, educated, enlightened global culture. It's now a small, bleak town, having been overrun and drained by one colonial ruler after another. Founded by Alexander the Great in 331 B.C., it quickly gained cultural and intellectual importance, attracting trade and culture from all over the Mediterranean. Its great library burned to the ground when Rome invaded the city out of jealousy. Later it grew as an early center of Christianity and again attracted the wrath of the Roman Empire. In the seventh century, triumphant Muslim forces established the new Egyptian capital up the Nile in Cairo. Alexandria slipped into quiet disrepair. Napoleon conquered Alexandria in 1798 and reinvested in it, recognizing its strategic value. When Egyptian leader Mohammed Ali took it back, he invested even more, making it one of the most vibrant cities of the nineteenth century. When the socialist dictator Gamal Abdel Nasser came to power in 1952, Europeans fled and took all their money with them.

But modern Egypt never forgets. The government of President Hosni Mubarak has established a great library in Alexandria once again. It has collections of ancient scrolls and books. It also has two hundred Hewlett-Packard computers running an archive of almost every page that has appeared on the Internet since 1986. The new library is part of the Egyptian government's effort to position itself among the leaders of the information age. It justifies massive investment in technological infrastructure by hoping to attract capital from abroad and prevent trained engineers from leaving for Europe and North America. The new library at Alexandria symbolizes the great hopes for Egypt's future. But it is not a real library for Egyptians. It's a tourist site, a mark of pride, but it's not a center of exploration for an eleven-year-old child who came to Egypt with his family from the Sudan. It's not a community meeting place that fosters social capital. It's a shrine and a museum, but little more. Still, it's a powerful symbol of liberalism without democracy, sponsored by a nation-state with too little of either. Libraries are never as placid as they appear. They are often sources and centers of controversy and conflict. The better they are, the more dangerous libraries can seem.[3]

Libraries Under Suspicion

Five days after the 2001 terrorist attacks on the Pentagon and World Trade Center, the *Washington Post* reported that investigators were looking into the fact that some of the terrorists had used open Internet terminals in public libraries in Fairfax County, Virginia, and Broward County, Florida. Clearly public libraries could be used to send e-mail to anyone in the world, and they offered many other Internet services without restrictions or surveillance. These policies grew out of traditional principles of confidentiality and privacy for library users, and respect for the general spirit of open inquiry within the walls of American public libraries. Nonetheless, such openness is now seen as a luxury that a secure society can no longer afford.[4]

Five weeks after the attacks, the U.S. Congress passed, without debate, and the president signed into law, without hesitation or deliberation, a 342-page document that hardly anyone had read completely: the *Uniting and Strengthening America by Providing Appropriate Tools Required to Intercept and Obstruct Terrorism (USA Patriot) Act of 2001.* The act radically revised legal protections against government surveillance of electronic communication. It eased the burden on federal law enforcement agents who monitor Internet and telephone traffic. And it severely challenged librarians to adhere to their Enlightenment principles.[5]

At least 545 libraries reported tangling with law enforcement inquiries about patron reading habits and Internet use in the year after the attacks of September 11, 2001 (178 of these were from federal investigators). The actual number might be much higher because it does not include inquiries under the special provisions of the USA Patriot Act that apply to libraries and bookstores. Libraries can be forbidden from revealing that they have been contacted by federal law enforcement officials. The Patriot Act is self-denying: If you're investigated under its powers, you're not allowed to tell anybody that you're being investigated. If you run an institution like a library or a bookstore and the FBI comes and says, "We want a look at all the records of this particular patron," you're not allowed to complain about it, protest it, or even inform the person who's being investigated. You're bound to secrecy. In other words, you're enlisted in the world of security and law enforcement whether you want to be or not.[6]

We don't know the effects of the Patriot Act—and may never know. Congress doesn't know how many times the act has been invoked or how many people are being investigated under the new secretive provisions. Therefore we don't know how effective the act has been and we have absolutely no way of testing it. The Patriot Act is a blank check to a government institution—the Federal Bureau of Investigation—that is notorious for overstepping its bounds, being ineffective, incompetent, and racist. This is not the sort of power we should give to any government agency without very careful oversight. But that's exactly what Congress did.[7]

Libraries have become sites of conflict in the new security state, perceived as dens of terrorists and pornographers. This is a poor description of how libraries work, as well as a dangerous assumption. Librarians are being forced to choose among their values. Librarians believe strongly in keeping records and maintaining archives, and library records are part of the historical record. But they also serve and protect their patrons. The federal government has forced librarians to choose between retaining records that might be useful (e.g., budgetary discussions or historical research) and protecting their patrons, who are not supposed to be intimidated by potential oversight of the books they choose to read. Many librarians around the country are taking a noble stand against this intrusion into their patrons' privacy.

Librarians should be our heroes. The library is not just functionally important to communities all over the world; it embodies Enlightenment values in the best sense. A library is a temple devoted to the antielitist notion that knowledge should be cheap if not free—doors should be open. Supporting libraries—monetarily, spiritually, intellectually, and legally—is one of the best things we can do for the life we hope to build for the rest of the century. It's no wonder that Fidel Castro chose to arrest librarians in 2003 while the United States looked elsewhere.

Washington's rush to generate a false sense of security is not the only threat to libraries. Hollywood is threatening them from the other coast. Hollywood dreams of a perfectly efficient distribution system, in contrast to its current inefficiency. Studios invest millions of dollars in products that may bomb on the market. They have no way to steer these ocean liners deftly through production process and promotion. Businesspeople in Hollywood seem constantly nervous because they never know what's going to win or what's going to lose. They don't know what their markets and audiences really want; they don't know how to adjust things midstream. Despite constant pressure to make their systems more predictable, creative products like films are unique and the markets for them are inherently unstable.

The pay-per-view notion comes from the desire for a more efficient, predictable distribution system. In a pay-per-view system you're not paying for thousands of prints of a movie; you're paying to keep the digital material on a handful of servers. The people who tap into this server are precisely the people who want to watch it, and if they're charged low prices, they don't feel ripped off. But installing this kind of pay-per-view system (much like cable TV) in all forms of culture—building a global jukebox—requires controlling every step of the commercial process in terms of format and content. Building the global jukebox first required passing the DMCA and then enforcing it vigorously. The oligarchs fear that we may build an open jukebox by ourselves with our own material. Unfortunately for them, that's exactly what we're doing.

Libraries are under incredible pressure to conform to the pay-per-view model. Increasingly, academic journals are coming to libraries in electronic form rather than on paper. So imagine this: An electronic journal is streamed into a library. A library never has it on its shelf, never owns a paper copy, can't archive it for posterity. Its patrons can access the material and maybe print it, maybe not. But if the subscription runs out, if the library loses funding and has to cancel that subscription, or if the company itself goes out of business, all the material is gone. The library has no trace of what it bought: no record, no archive. It's lost entirely. This is not a good model for a library. It defeats the fundamental purpose of a library. You might as well be sitting at a computer terminal in a copy shop.

The Perfect Library

Perhaps Washington, Havana, and Hollywood pressure libraries because they overestimate the openness, convenience, and power of the information flows within them. Libraries are not good enough at connecting us, protecting our privacy, enabling access and copying, and maintaining archives to threaten the powerful in American society. The roots of this overreaction may lie in the specter of stronger, fuller, more powerful information flows that new technology might enable: a perfect library.

Imagine the perfect library. It would offer you access to any text, song, film, image, or video game. It would be easy, convenient, and free. Its indexing system would allow you to search for and find the smallest piece of information. You could use any number of books without denying anyone else access to them. You could do most of your searching from your home or office. The marginal price of using library services would be zero. Users' privacy would be ensured, so no power could frighten or embarrass people from going wherever their curiosity led them. The perfect library would be built and stocked by volunteers who donated their time, labor, creativity, and passion. Users would donate their personal libraries and volumes of fresh content. The perfect library would be funded at first by various government entities, but governments would back away from that role and would not regulate its use in any way. The veracity and utility of the information available through the perfect library would vary, but users would supply communally edited analyses that would rate and rank them. There would be enough outlets for everyone to have reasonable access regardless of class or wealth. The doors would be open at all hours, the lights always on. The perfect library would be more than a repository for information. It would be a communication medium as well.

The perfect library might have some powerful positive effects on the world. There would be no information monopolies. Everyone would have equal access to facts and poems, techniques and tirades. Citizens of all nations could test their governments' claims against other sources. If the perfect library's indexing system became the main source of what we want to know about the world, CNN or Fox would have no advantage over small newspapers in Ghana or radio stations in Quebec. We might live in a world with diversity of thought and culture, a true free market of ideas. The perfect library could be a powerful resource for the expansion and enrichment of democracy.

On the other hand, the perfect library would be a haven for those who wished to abuse these freedoms to harm other people. Child pornographers could trade their materials without fear of shame or prosecution.

Pranksters could exploit the openness of the system to disturb or impede innocent users. The industries that produced the library's vast holdings of books, songs, and films might wither or collapse because nobody would want to pay for cultural products or information. Absent incentives for investment in large productions or aggregations, all new content would be low-budget, handmade expression. There might be no more *Star Wars* episodes from George Lucas. There would be no new *Encyclopedia Britannica*. Filtering the true from the false and the good from the trashy would require constant vigilance and refined skills. Most alarming, criminals and terrorists could use the perfect library to speak in code, access sensitive documents and schematics, and research new methods of killing large numbers of people quickly. In this way, the library would be "perfect" as in *Perfect Storm*.

Both visions of the perfect library—utopian and dystopian—are overstated. We are not close to constructing the perfect library, but we can imagine how it might look and act. Many of our communal efforts since the early 1990s seem to be moving our information ecosystem toward that vision. Yet long before we get there, many are sounding alarms about the ways people might abuse their freedoms to use and move information. Even though the perfect library is not imminent, many are acting as if it is. The strong reactions of those who would squelch these freedoms might render our information systems unable to perform the positive functions of the perfect library because of unexamined—often merely assumed—threats to the status quo. The closer we get to the perfect library the more oligarchs try to undermine it.

Harsh reactions to information anarchy in recent decades include technological access restrictions, electronic surveillance measures, coercive contracts, stiff legal penalties for distributing information without authorization. Strategies of propaganda, confusion, and distraction generate a pervasive chilling effect on the public. These reactions are overkill and reach far beyond the communications networks themselves to corrupt the inner workings of culture, science, education, commercial competition, and even democracy itself. Rather than the perfect library, the information system being imposed resembles a pay-per-view universe in which all terms of use and reuse are dictated by the information provider and licensed by the state. The enemy may be the specter of the perfect library, but the victim is the real library.

The Real Library

A Saturday afternoon at the Brooklyn Public Library offers hope. Children from four to fourteen gather around desks, reading books, scribbling in notebooks. A devoted staff rushes to reshelve material. Librarians answer questions with smiles and patience. No one seems rushed or hushed. Everyone is respectful and calm. The Brooklyn library, a classic example of American Works Progress Administration modernism, sits at the nexus of several neighborhoods and attracts patrons from some of the wealthiest and poorest sections of the borough. A quick stroll from the Botanical Gardens, Prospect Park, and the Brooklyn Museum of Art, the library is the intellectual, cultural, and civic heart of Brooklyn, which is a cosmopolitan and global city unto itself, not so different in tempo and timber from Cairo. Just as Cairo is the center of Islamic culture in the Middle East, Brooklyn is a center of Islamic culture in North America. The community clearly takes great pride in its library. It's more than a functional source of knowledge and peace in an otherwise noisy and hurried borough. The Brooklyn library is a statement that the city cares enough to build and maintain such a functional monument to the Enlightenment. There are libraries like Brooklyn's all over the United States. And many communities use them with just as much enthusiasm. In 2004 there are more libraries than McDonald's restaurants in the United States. Many millions are served.[8]

Libraries are a threat to the content industries and their plans for a pay-per-view delivery system. Libraries are leaks in the information economy. As a state-funded institution that enables efficient distribution of texts and information to people who can't afford to get it commercially, the library pokes holes in the commercial information system. Because a library can lend a book at no charge, the publisher only makes money once. It can't charge per reading. The new technocratic information regime aims to correct for that market failure by regulating access. If books became streams of data rather than objects for sale, they could be metered, rendering libraries superfluous or relegating them to vendor status. There would be nothing "public" about them. What we now think of as a library—a solid building full of books, ample tables, comfortable reading chairs—might look more like a modern office. It might be a storefront with rows of carrels or cubicles, each with a computer terminal and printer. A patron would enter a credit card or debit card to access databases of text, music, video, or facts. The computer would charge the card by the minute or the megabyte. It would pour the content onto a disk, stream it to your personal computer or account, or print it out on paper. There would be no functional difference between your neighborhood library and a Kinko's or Barnes and Noble Superstore.

The emerging information pay-per-view regime could signal the death of the liberal Enlightenment project and thus the public library itself. Thinkers as politically diverse as Theodore Adorno, Gertrude Himmelfarb, Dinesh D'Souza, and Neil Postman have predicted the death of the Enlightenment, blaming a variety of causes, including the Enlightenment's internal contradictions, the empty cleverness of postmodernism, and the social tremors caused by ethnic politics. The real culprit, however, is the steady commercialization of the cultural and communicative process.

There is a real value in a public library and its metaphorical counterpart, the information commons. Though not quantifiable, it is discernible and essential. Public libraries are functional expressions of Enlightenment principles. We are about to let commercial interests shut them down (this will be news to most people but not to librarians). The public library is where the public domain lives, the place where we gain access to the information commons. As our rights of fair use, first sale, and use of works in the public domain disappear, so do the raw materials of our culture and democracy. Most importantly, the public library is where those without money, power, access, university affiliation, or advanced degrees can get information for free. Thus trends toward a pay-per-view delivery system threaten both the public library and Enlightenment ideals. They signal the dawn of the age of proprietary information.

An open global information ecosystem is essential to a dynamic culture and the spread of stable democracy. Our information environment is under attack on technological, legal, and commercial fronts. Information, far from being a scarce resource, is more abundant than ever. For centuries, our species suffered because of maldistribution of information, not necessarily an overall shortage or a deficient quality of information. Facing an apparent oversupply of information, we attempt to employ technological tools to correct the maldistribution of information. A child in South Africa might soon be able to read the same books and magazines I do—as soon as I do—at relatively low cost. We might be poised to celebrate what media scholar Mitchell Stephens has called the "End of the Era of Insufficient Information." Ideally, if we could distribute information access more equitably, we could make a good attempt at fixing many of humanity's ills. Information would no longer be power because power demands a difference—a maldistribution. If some nations did not enjoy an information glut and others an information deficit, perhaps the strong nations would have a harder time dominating the weak. Perhaps leaders could distribute food and fuel better if information about needs, demands, supplies, preferences, and efficiencies flowed freely. In the absence of good information, consumers, voters, and soldiers all make bad choices. Conversely, with good information, good decisions are at least possible. The sum of human decisions might be much better with cheap, abundant, high-quality information. At least we can assume that the sum of human decisions cannot improve without it.

Some people—Bill Clinton and cyberauthority Nicholas Negroponte included—have argued that we are just years away from the ability to connect every person on Earth to a vast library of knowledge, cultural expressions, and provocative ideas through satellite transmissions, phone wires, computers, television, radio, and print. They argue that once we work out the rules of engagement of cross-cultural discourse, we will be a lot closer to improving life everywhere. However, that techno-utopian vision is not emerging as predicted. The obstruction is not technological, religious, or social—but commercial. For information to be commercially valuable it must not be widely available. Content industries have an interest in creating artificial scarcity by whatever legal and technological means they have at their disposal. Conversely, citizens and consumers have an interest in abundant information. To be democratically, artistically, and scientifically useful, information must be cheap, bountiful, and accessible.

What if we looked out the window one day to find that someone had run a fence through our information commons? What if the raw materials for composition and communication were all locked behind private gates? We would have no Rolling Stones without the commons of Mississippi Delta blues. We would have no *Adventures of Huckleberry Finn* without the commons of American oral storytelling. We would have no informed debates about the value of social security or the designated hitter without the commons of widely available, dependable information.

The Information Elite

In 1956, C. Wright Mills wrote, referring to the political economy of information, "As the means of information and power are centralized, some men come to occupy positions in American society from which they can look down upon, so to speak, and by their decisions mightily affect, the everyday worlds of ordinary men and women." Mills was complaining about a phenomenon that undermined the principles of a democratic republic at a time when the United States should have felt most secure about its systems. The powerful tend to hoard information to maintain their advantages—the ability to react effectively to changes in society or the economy and the power to manipulate public opinion. Information discrepancies have marked societies since the first dominant leader insisted on keeping secrets from the general population. In Genesis, the serpent exploits universal human passions like curiosity and hubris, and consequently Adam and Eve are cast out of blissful ignorance into a world of horrible truths. God protects his information monopoly. The distributive effects of the printing press and transportation networks enabled the Protestant Reformation in Europe and the eventual rise of republics around the world. In each case, powerful, established forces—Church and Crown—tried to rein in the disruptive power of the new technologies while harnessing their benefits.[9]

While information discrepancies are omnipresent and in many cases essential, they should be limited in democracies. With reasonable and malleable discrepancies, trustworthy filters, and effective cultural structures and forums for reasonable exchange of and debate over information, a democratic republic can thrive. When high barriers limit access to information, only the elite participate effectively in the workings of a republic. At the beginning of the twenty-first century, information discrepancies have taken on new importance. They are about more than the ability to interpret information to gain a competitive advantage over others. Information itself has value. More than a raw material, it's a product—the main product of the most dynamic sectors of the global economy: entertainment, news, computer software, and various services such as consulting and planning. Genetic codes, encryption algorithms, and financial data are now expressible and tradable as information. Thanks to global electronic networks, these information products and services are infinitely replicable and distributable with no marginal cost. To maintain the market value of information products, elites must find ways to create artificial scarcity and restrict flows of information. But information is also an essential public good. As Amartya Sen has shown, famines occur not because there is too little food available to an afflicted population but because some nation-states restrict the feedback mechanisms of democratic accountability that would otherwise generate a reasonable reaction to natural disasters and maldistribution of resources.[10]

The Power Anxious

At a public symposium in 2000 celebrating a new edition of C. Wright Mills's 1956 masterpiece, *The Power Elite,* legendary activist Tom Hayden called for someone to write a new version of the book. Hayden credits Mills's work with inspiring his 1962 Port Huron Statement. Thus it contributed to the blueprint for the rise of radical democratic activism through the rest of the 1960s. Along the way, activism splintered and spun into many different orbits, some positive, some disturbing. Several decades later, the world is still governed by a cadre of powerful men who monopolize power over and above the naked skeletons of dysfunctional democracy.[11]

Do the same structures continue to exist, the same sorts of sealed communities of power and influence that existed in 1956? I don't think so. The most powerful world leader in 2004 is the son of a former president, grandson of a senator, graduate of Yale, and heir to a set of influential associates, five of whom selected him to be president over the expressed wishes of a majority of American voters. Yet much has changed. What would an updated power elite look like? The power elite is still powerful and elite, but it's very different. It's increasingly represented by impersonal institutions, structures, and technologies, rather than a distinct class or set of individuals. It's global and getting more so. The elite are now members of a multinational stratum of political, military, and corporate leaders. Their power is spread among a vast array of interests, many of which compete violently. Membership is more open, yet the price of being left out is much higher. Despite all its success, the power elite is vulnerable. In the world of constant communications, the channels have proliferated exponentially and each company's share of the public attention span has withered. Ownership is more concentrated under few corporate umbrellas, but corporations are increasingly owned by absentee, mute shareholders who are globally dispersed and kept ignorant. Shareholders are at least once removed from the actual governance of corporations by the institutions that aggregate capital. The owners of a corporation, unable to participate in governance, issue simple signals to corporate boards and officers. They push the managerial elite to produce higher returns over shorter periods. Richer feedback between owners and managers might yield more subtle, stable, long-term growth, but the communication dynamics of corporations do not allow for that. So the managers lunge at any new business model consultants happen to imagine. Managers have been monetizing processes and resources—most importantly elements of intellectual property—with a diligence never before seen. Many of these business models embrace radical free flows of information. Others embrace radical restrictions on the use and flows of information downstream, among users, consumers, and citizens. Firms have been redesigning their delivery and revenue systems haphazardly and counterproductively. The power elite are now the power anxious, frightened of diminishing market power in an unpredictable global mediascape. Mills diagnosed a malady for his age that has grown more malignant in ours: the dangers posed by those who seek to centralize and concentrate information.[12]

Out of Balance

A trenchant insight in *The Power Elite* was that by the 1950s the Madisonian balance on which the American republic traditionally depended no longer operated. Too often, Mills argued, interests that were supposed to compete and mitigate each other instead acted in tandem, so that a civilian Congress could no longer curb the excesses of a professional military and its industrial partners. State and private interests no longer had distinct spheres or concerns, and private parties increasingly aligned themselves to rig the outcome of political games. In this way they maximized the output for elites at the expense of the public. Mills underestimated the extent to which the state still chooses between guns and butter and mediates between starkly divergent visions of the good life. But he did alert us to the trend of political trusts operating much as economic trusts did during the late nineteenth century. Today, such unseemly alignments between states and corporate interests appear in the body of such global governing institutions as the World Trade Organization, the World Intellectual Property Organization, and the International Monetary Fund. While these undemocratic institutions assume more control over the daily lives and policies of sovereign nations, the global resistance to these powers increasingly operates along distributed models.

We need a just, reasonable, republican model of information distribution. We should be able to enjoy and exploit the freedom of information anarchy with the ability to discern good from bad, useful from useless information. This will require time and patience. Mostly it will demand cool, sober deliberation about what ethical, legal, and technological limits appropriately regulate the flows of information. Although we need to generate a sense of global information justice and global cultural justice, we have bickered over the specifics in narrow battles over use and abuse of material. We are a long way from developing such a sense of justice and the cultural habits that might empower it. I fear we may be too late.

CHAPTER 9

The Anarchy and Oligarchy of Science and Math

During the Cold War, scientists behind the Iron Curtain yearned for life in the United States. Basic needs and conveniences were better met in the free world, and the principles of open dialogue and frank examination created fulfilling intellectual communities. Scientists were among the few Soviet citizens allowed to travel frequently to western Europe, North America, and India, and consequently they were among the first to see through the lies and exaggeration of Soviet tyranny. In early 2001 Russian scientist Elena Bonner gave a speech about the lurch toward authoritarianism in Russia under President Vladimir Putin. She stated that if not for Soviet scientists in the 1960s, dissidents would have had no sense of the shell of lies in which the government had encased Soviet society. Soviet scientists, having communicated with the outside world, let a little light and a little air into an otherwise blind and suffocating nation. They were led to do so by the ideology of their profession.[1]

Science is the most successful open and distributed communicative system human beings have ever created. The cultural norms of science, and by extension academia in general, are ideally cynical and anarchistic, and science and academia should be radically democratic. Although membership in these communities is limited to a select few, the papers and books that the members produce are usually open to public perusal and commentary. And the traditions of blind peer review allow motivated amateurs to participate occasionally in discourse and discovery, even if they can't get past the guards protecting labs and libraries.[2]

Science is a culture and a method. It embodies an ideology that supports the method and maintains the culture. And it's linked to a set of industries through which billions of dollars, public and private, flow every year. The stakes of science have never been higher nor the justifications for it clearer. World War II, we are told, was won because one side had a group of well-funded immigrant scientists who developed better technologies, including radar and code-breaking methods, than the other side did. The losing side, having purged most of its top scientists for the sake of ethnic and ideological purity, failed to meet important technical challenges. Ultimately the side that allowed scientists to work with confidence developed a better bomb as well. The challenges of the twenty-first century—poverty, security, and disease—may all be addressed with ideas that start in the laboratory or computer and flow out to the market, the farm, the school, or the clinic. Healthy, uncorrupted scientific communication is a necessary (albeit insufficient) condition for improving the human experience.

Scientific knowledge often moves from a spring of open discourse into a stream of adoption and exploitation and from the public arena to the private sector. Complex protocols guide this process, each step embodying different values and ideologies. The rules and terms of discussion begin with consensus-seeking processes within scientific communities. They then consider the demands of market forces to create and enforce scarcity and state demands for security. Different ideologies, habits, and rules govern the "upstream" source of knowledge and the "downstream" deployment of it. But the first step, the action in the lab and the library, depends on academics' devotion to radical democracy and openness. The essential question in this matrix of rules and norms is, At what point in the knowledge stream should we install controls and restrict access—to encourage new technologies and to protect people from bad actors who would exploit dangerous knowledge?

Members of scientific communities face significant real-world barriers to true and ideal openness and equality. The first is the relatively soft barrier of expertise. The rare amateur in theoretical physics must spend years mastering the body of work that preceded her curiosity. Without such mastery and the time to pursue it, a potential contributor would not know the existing gaps in the knowledge, the particularly interesting questions, or blind pathways that had already been scouted. Although scientific discourse is ideally open to anyone with the requisite knowledge and expertise, becoming an expert demands a significant investment of time and money. This process tempers the potential excesses of information anarchy: the persistence of rumor and error and the cult of personality.

The second, harder barrier is credentials. In a messy, crowded, busy world, degrees and titles serve as imperfect proxies for knowledge and expertise. You might not know whether it's worth your time listening to a discourse on the virtues of genetic engineering given by the person seated next to you on the train. But if she introduces herself as a professor of molecular biology at Rockefeller University, you might decide to listen. Credentialism is inherently oligarchic. Admission to the academy of credentials is severely restricted, as its members prefer to limit competition for jobs and resources. Credentialism can be self-perpetuating. A board of credentialed experts reviewing grant applications is likely to dismiss applicants who lack the same basic credentials they have earned and reward those who went to the right schools, regardless of more subtle measures of knowledge or expertise. Credentialism incorporates all the potential excesses of oligarchy. The professor on the train could be full of hot air, as many professors are. The chief problem with credentialism comes from the synergy of status anxiety and arrogance: such professionals might be less willing to admit error than an amateur or novice might. Fortunately for scientific progress, any group of credentialed experts is likely to generate significant disagreement on the burning questions of the day. So expertise trumps credentialism and real debate can occur. It's impossible to know which conversations and debates don't happen because of the inherent conservatism among communities of the credentialed. Despite some elements of oligarchy, science as a practice succeeds because of, not despite, its ideology of relative openness. Credentialism is more an imperfection than a corruption of science.

As an ideology and culture, science is supposed to be open to contributions from the unlicensed. Unlike the humanities, where credentialism is a much bigger problem, science can be somewhat free from the tyranny of credentials. It's supposed to be unconcerned with questions of nationalism or commercial gain. Although the public hails legends like Isaac Newton and Albert Einstein who have broken open scientific fields and rewritten textbooks, science is most often done by teams of researchers, collaborating in even larger communities across borders and oceans. Science has always been global, cosmopolitan, messy, inefficient, and troublesome. With the rise of global communicative technologies and more sophisticated methods of computer modeling in areas as diverse as cell biology and nuclear physics, the barriers of entry should be lower than ever and collaboration and criticism should be easier and cheaper than ever.

One community of researchers and creators—the free software movement—has adopted radically democratic academic principles as its guiding philosophy. While professional and degreed computer scientists make significant and notable contributions to the evolution of free software, the amateur matters greatly. Often it's amateurs who debug and improve a piece of code or find a way of using it in a new context. Computer science is new enough and its tools are cheap enough that thousands of amateurs who lack credentials gain expertise through trial, error, experimentation, collaboration, and communication. It's a scientific community that Francis Bacon would have envied and Aristotle could not have imagined. Recently it has emerged as a placeholding metaphor for values and habits that have much older currency. Free software (often conflated with the ideologically neutral term "open source") has become a model and an argument, yet its principles used to be unarticulated because they were the default in science.

Just as communication technology allowed the flowering of a new scientific revolution, the oligarchic concerns of commerce and national security crowded out democratic values at their source—the university and laboratory. More than a decade ago Elena Bonner and her husband, Andrei Sakharov, helped end the Cold War by spreading the ideals of scientific openness to the entire Soviet Union. Now we must start questioning whether the United States will be a haven of science in the future. Many scientists and mathematicians, citing legal threats against encryption researchers and the criminal prosecution of Russian computer scientist Dmitry Sklyarov and nuclear scientist Wen Ho Lee, as well as increasingly strict visa restrictions governing students and researchers, have been frightened away from visiting or working in the United States.[3]

American scientists are finding it harder to do their jobs openly in the new security environment since September 11, 2001, and the as yet unsolved anthrax attacks that followed. Over the past few years, the U.S. government has severed important links on federal World Wide Web sites, deleted information from other government Web sites, and even required librarians to destroy a CD-ROM on public water supplies. University of Michigan researchers lost access to an Environmental Protection Agency database with information they were using to study hazardous waste facilities. Unclassified technical reports have disappeared from the Los Alamos National Laboratory Web site.[4]

Rules regulating the use of dangerous materials or the distribution of information that could be abused evolve slowly through the scientific process. Groups of scientists, in concert with government officials, examine risks and propose restrictive protocols. Some are encoded in law, while others remain part of the self-regulating culture of science. But since 2001, the U.S. government has taken to dictating the new security rules, regardless of the scientific merit of the restrictions. Many of these rules have generated criticism among scientists who fear a chill on certain essential research (e.g., on bioterrorism) and on the review process that requires other researchers to replicate previous experiments. If some data or conclusions are kept secret, science cannot proceed in a self-correcting fashion. Most alarming, the U.S. government has decided to restrict and monitor contacts with non-U.S. scientists and graduate students. The global, cosmopolitan nature of science is threatened if the world's largest source of basic research explicitly favors its own citizens instead of letting the best American scientists participate collaboratively and internationally with the best non-American scientists.[5]

Even before the attacks of 2001, something serious was changing in the relationship between science and the U.S. government. Since the early 1980s, university administrators and political leaders have put increasing emphasis on the potential profitability of publicly funded basic research. This, coupled with widespread concern for the perceived security risks of open networks, open journals, and open discussion, has pushed scientists to reassert their principles and defend their peers. Battles have erupted over the content of journal articles, the control that journal publishers exercise over material, the role of foreign-born and ethnically suspect scientists, and the ethics of privatizing basic information about the world and the human body. Scientists are fighting for the Enlightenment all over again.

Commerce and Control

As molecular biologist Roger Tatoud has written, "It is widely accepted that science should be an open field of knowledge and that communication between scientists is crucial to its progress. In practice, however, everything seems to be done to restrict access to scientific information and to promote commercial profit over intellectual benefits."[6]

Tatoud is most concerned with the increasing influence of two systems of regulation on the culture of science: copyrights and patents. Copyrights directly affect the price of scientific journals and thus their availability to researchers in developing nations and at poorer institutions, as well as those unaffiliated with a company or university. The absurd copyright economy forces scientists to assign all rights to a major commercial journal publisher for no remuneration, then buy back the work through monopolistic subscriptions. As a result, many scientists are forming free and open collaborations to distribute peer-reviewed scientific literature outside the traditional commercial journal system. The Gordon and Betty Moore Foundation is sponsoring the "public library of science" and the George Soros Foundation funds the Budapest Open Access Initiative. The Web site for the Budapest project describes it as the convergence of "an old tradition and a new technology . . . to make possible an unprecedented public good. The old tradition is the willingness of scientists and scholars to publish the fruits of their research in scholarly journals without payment, for the sake of inquiry and knowledge. The new technology is the Internet. The public good they make possible is the world-wide electronic distribution of the peer-reviewed journal literature and completely free and unrestricted access to it by all scientists, scholars, teachers, students, and other curious minds." The initiative's goals include plans to "accelerate research, enrich education, share the learning of the rich with the poor and the poor with the rich, make this literature as useful as it can be, and lay the foundation for uniting humanity in a common intellectual conversation and quest for knowledge."[7]

While the copyright system benefits the publishing oligarchs at the expense of scientific openness, it has not restricted science as much as the patent system has. In 1980 the U.S. Congress passed the Bayh-Dole Act, which encourages universities to patent work generated with public funds, and the U.S. Patent Office approved the patenting of living things and the genes that operate in them. Since then there has been a mad rush to control information that might be medically relevant.[8]

Myriad Genetics, Inc., is an American company that managed to wrest control of two mutant genes that influence breast cancer in a small number of women. It has collected immense monopoly rents from medical care providers who must pay the company $2,500 each time they screen a woman for these mutations. As British biologist John Sulston has written, "By claiming proprietary rights to the diagnostic tests for the two BRCA genes and charging for the tests Myriad is adding to total health-care costs. Even worse, once scientists really understand how the BRCA 1 and 2 mutations cause tumors to grow, they might be able to devise new therapies. But because of these patents, Myriad has exclusive marketing rights." In other words, researchers have a financial disincentive to act as free agents when developing new tests and therapies for these mutations. And throughout the world, these tests remain beyond the financial reach of billions of women.[9]

The triumph of centralized information control over openness and efficient short-term commercial gain over the long-term accumulation of knowledge is the major theme of this story, but it's not the only one. In many of the battles between openness and control of processes and information, overcontrol has had a perverse effect on commerce. Proprietary control of databases of essential genetic information, for instance, raised the specter of redundant, imperfect, competitive private databases simultaneously lowering profits for companies that maintain them and raising transaction costs for companies that wish to use the information to develop drugs or therapies. For this reason, several pharmaceutical companies have joined with the Wellcome Trust in the United Kingdom to form a free public database of SNPs (single nucleotide polymorphisms), a common DNA sequence variation among individuals that results from a single "letter" difference in the code. By identifying the SNPs, researchers might identify susceptibility to specific diseases that have genetic influences. Before the public SNP database obviated the "gold rush" to identify and patent hundreds of SNPs, lone companies were trying to hoard the information and patent the SNPs. Had they succeeded, research on particular SNPs would have been more expensive and potentially monopolistic. The public SNP database is a case in which companies heavily invested in a healthy and reliable patent system overtly avoided abuse of the system by investing in public domain information. They realized that too much control by too few was bad for business as well as science.[10]

The U.S. government had nothing to do with the open public database, besides funding some SNP research. U.S. science policies heavily encourage universities, public sector researchers, and private companies to file for patent protection on every step of the knowledge-producing process. These policies have generated an exponential increase in the number of patents owned by universities for work done with public funds. In 1979 American universities received 264 patents. By 1997, that number had increased tenfold, to 2,436. In that same time, the total number of U.S. patents issued per year only doubled. U.S. science policies have erased any functional difference between the ways universities regulate and license basic science versus commercially exploitable technology. Perhaps most importantly, the American people are paying at least twice for any research that generates a marketable technology or treatment—through the federal grant and through the market price of the procedure or drug.[11]

Pardon Me, May I Use My Genes?

Despite its success, the lessons of the open SNP database were lost on some. In early 2000 the two most prestigious, widely read scientific journals in the world, *Science* and *Nature,* published competing articles that outlined revisions of the human genome. Unlike its British counterpart, *Nature,* the editors of *Science* decided to bow to the demands of the private genetic research company Celera. The actual genome data cited in the *Science* article is available only from the Celera Web site. But to get access to the Celera data, you must agree to a set of highly restrictive rules. To see what is nominally "published" data, you must be eighteen years old, be affiliated with an academic or nonprofit research institution, and provide Celera with substantial background information. Once approved by Celera, scientists must promise not to redistribute the sequence or commercialize any procedures they develop from the data. *Nature,* in contrast, did what scientific journals are supposed to do. It openly and freely published the results from the publicly funded National Human Genome Research Institute in their entirety. Anyone may read and use the data without asking permission or agreeing to any restrictions. The data from both sources is similar, and most scientists working with this information consult both databases for annotations and details.[12]

Information, like the homeostasis that maintains and in fact defines life, is a struggle against entropy. A healthy information system maintains a general equilibrium while allowing for change, adjustments, improvisations. Ideal systems are open, flexible, decentralized, and rich with diversity and differentiation. Weak systems are rigid, closed, exclusionary, inflexible.[13] An example of a frustratingly open, almost completely unregulated information system is the Internet. An example of a regulated system that has specific features meant to encourage openness and flow would be a university. A more closed system might be a high school. An example of a highly regulated system would be cable television. The project, process, and community that we call science is partially regulated yet intentionally open. One generally must pass through an extreme licensing system to enter the community, but independent thinking and skepticism are encouraged among those inside. Claims are tested and verified through publication, communication, and experience. These are the ideals; the practice is inevitably flawed, but aspirations to openness give the community a basis for criticism and positively influence the process. Much of the debate in scientific ethics involves this dialectic between inclusiveness and exclusiveness, between convention and imagination, and between ideal and reality. One tenet of the scientific ethical canon is that scientific work should be open to examination, criticism, and use by others. Thus a highly regulated information system is contrary to honest and effective science.

Still, science for the sake of science is only part of the story. The application of knowledge toward particular technologies demands an incentive system. Without a dependable expectation of return on a risk or investment, companies and individuals might not strive to invent useful things. In other words, technological advances are subject to market failure. If an innovator cannot restrict the use or application of a technology or cannot create scarcity for a product, she can't charge monopoly rents for it. Without temporary, limited monopoly power, we might not enjoy many of the useful technologies we have.

In April 1992, James Watson, one of the discoverers of the structure of the DNA molecule, resigned from the National Center for Human Genome Research. He did not quit because the project was failing or because of some scandal or personal strife. He resigned because he opposed a major step in the project. Craig Venter, founder of Celera, and the directors of the National Institutes of Health insisted on pursuing patents on the genome. When the project leaders were just firing up for the massive goal of determining the sequence of what they thought was going to be

50,000 to 100,000 human genes, no one was more enthusiastic about its possibilities than Watson.[14]

The Genome Game

Genes are segments of chromosomes that encode the building of proteins; each protein functions in the structure and regulation of the cell and therefore the entire body. Knowing which gene influences which function was supposed to be the key to predicting and preventing genetically influenced diseases. Each gene is composed of bundles of deoxyribonucleic acids (DNA), the detailed blueprints of life. DNA is composed of nucleotides, building blocks differentiated by one of four sugar molecules: cytosine (C), guanine (G), adenine (A), and thymine (T). Three nucleotides in a row (known as a codon) code for a specific amino acid, a protein building block. The order of all body proteins is determined by the order of DNA codons. The lengths of DNA are called chromosomes, and the normal human cell has two sets of twenty-three homologous chromosomes—one set from each parent. Chromosomes are often visible through a light microscope in the nucleus of a cell. The genome is the macroview of the series of genes—the entire assemblage. The goal of the Human Genome Project was to assess each gene's location on the chromosomes, locate and distinguish among all the different components of the genes, establish their functions and interactions, and eventually understand how they work together to create and maintain cells, systems, and organisms as a whole.[15]

The Human Genome Project was a huge multinational effort funded largely by the public and guided by the NIH. The project had two defined objectives: To create a map showing the location of each gene and to determine the order of all the DNA in each gene.[16] The U.S. government committed more than $3 billion to the effort, divided among nine laboratories across the country.[17] Some of the American research was also funded by a consortium of pension and insurance companies called the Healthcare Investment Corporation. The Japanese government funded its central genome research. The French biotech company Genethon funded the French lab work with money raised from telethons to defeat muscular dystrophy.[18]

After the public uproar over the NIH plan to patent the genome, and the subsequent rejection of its application, the question of patenting the bits of genes that the NIH produced seems to have been be answered. But some issues that transcended the NIH patent applications remain unresolved. First, what financial reward should be available to those who toil to compile data that has no immediate use? Second, if we reward those who amass this information (and other such databases), how do we ensure that an entrepreneur can afford to use the information, as well as profit from it? Third, how can we encourage such research without squelching future research and the altruistic research projects that might spring from it?

Several experts in this field suggested changes to the patent process. In the early 1990s Craig Venter suggested amending the patent law so that a previously patented fragment of a gene would not render the complete gene "obvious."[19] Such action might neutralize the concerns of many industry leaders, but it simply inserts a loophole in two centuries of law that generally handles standard inventions effectively. It is a political solution that might answer the needs of this specific industry. But what about the next industry that relies on publicly funded data compilation? Would Congress then insert yet another loophole? Some, including an editorial writer at the *Economist,* suggested a move in the opposite direction: clarifying and tightening the usefulness test for patents, making it harder to get them.[20] Others suggested that a system like copyright might be simpler and more affordable.[21] Actually, a copyright-like system would be useless. Any researcher who copyrighted his or her data set would enjoy insufficient protection. The researcher would receive credit but not much else. Copyrights do not protect ideas, just specific expressions of ideas. With much scientific data, and especially genetic codes, the idea is the expression.[22]

A compulsory licensing system might be applicable to an array of controversial intellectual property issues such as music sampling and databases. The federal government or some international agency would sponsor a grand genetic database in which all genomic information could be stored and compared (and weeded for errors). The researcher would be paid a fee by the codon (or by the gene) for his or her submission to the genome database. The company that takes the raw material (the data) and forges the final profitable product (the drug) would pay a reasonable fee to the nonprofit library. But no one could be excluded from examining the information in the protected data set. A company could properly patent the process that led from the genetic information to the final product. The trick, of course, is to balance the fees so that the company is not discouraged from trying out new treatments, yet the researcher is not discouraged from putting in years of effort on something that offers little immediate reward. If the compulsory license system worked, the company's interests would be protected, the researcher would be rewarded, and no one would own a patent on the genetic makeup of human beings.

Celera, in collusion with *Science* magazine, employed a system similar to the compulsory licensing scheme but with restrictions. Celera put its data behind a high fence. Anyone who desires access to the data must submit to a lengthy questionnaire, certify affiliation with a major nonprofit research institution, and agree to a lengthy user agreement. No one may use the information for profit without negotiating terms with Celera—Celera takes a piece of any patentable and licensable techniques that may arise from the research. And the information must not be redistributed publicly. This elaborate contract creates artificial scarcity. The information, digitally rendered, is restricted by promises and pledges and by an electronic gate. Celera can deny access to anyone. Databases are not yet protected by specific intellectual property law in United States, but they are in most of Europe.

There are severe problems with a state-granted monopoly system on treatments of any kind. Global flows of goods and capital have made differential pricing and differential information policies almost impossible to maintain. Many of the world's poor—and thus many of the world's sick—are shut out of the best and in some cases only treatments for the diseases that ail them. A pharmaceutical company that used Celera's research to treat sickle-cell anemia, a genetically carried disease, might be able to alleviate the suffering of millions of people in the tropics. But if the treatment were priced for the U.S. consumer market, where there are hundreds of thousands of sickle-cell sufferers, it might never reach the millions in the tropics who need it.[23]

The good news is that the success of the publicly funded and openly published genome project serves as a model for how science can and must be done. Its success speaks to the virtues of open information systems, collaboration, and the advantages of the application of information technology. Negative externalities—misuse and abuse of these technologies—are more likely in private, opaque environments.[24]

The genome project has intellectual implications that we are just starting to grasp. Openness makes biological complexity impossible to ignore. Research on the genome has revealed far fewer genes that anyone expected. Until recently scientists were predicting that humans had from 50,000 to 100,000 genes. And everyone assumed that a specific gene coded a specific protein. As it turns out, we have about 30,000 genes, just over twice the number needed to produce a fruit fly and just 50 percent more genes than a simple roundworm. How can such a small set of information code for an organism as large and complex as a human being? As Stephen Jay Gould explained in the *New York Times*, we now know that each gene produces several kinds of proteins under specific environmental conditions. "The collapse of the doctrine of one gene for one protein," Gould wrote, "and one direction of causal flow from basic codes to elaborate totality, marks the failure of reductionism for the complex system that we call biology." In other words, we can't assume that we are our genes and our genes make us. We also make ourselves. So do our schools, our parents, our weather patterns, our predators, our diets, our books, our houseplants, our decisions. We are not mere code. We are influenced by many things, determined by none. The immediate intellectual leap we can make from this simple arithmetic conclusion is that there is no place for eugenics in any vision of our genetic processes. Eugenics is much harder to imagine once we admit complexity, dynamism, and feedback mechanisms inherent in human development.[25]

Another implication of the 30,000 genes—each of which has more than one function—is that we can no longer justify protecting the data on a single gene if it produces multiple effects—perhaps multiple therapeutic effects. Limiting access to and use of information about our genes could cause severe harm and stifle exploration and invention. And such a policy would rest on false assumptions about how genes work.[26] As historian Evelyn Fox Keller explains in her book, *Century of the Gene,* the principles behind genome regulation—gene patenting and licensing—perpetuate one of the most dangerous myths in modern pseudoscience: the Jurassic Park fallacy. You cannot raise a dinosaur by cloning red blood cells taken from mosquitoes suspended in amber. Genes don't work that way. Organismic development is dynamic and complex, and it depends on a million environmental factors—not least the unique protoplasmic makeup of a zygote.

Differentiated adult cells cannot host genetic material in the same way. Red blood cells, especially, cannot do this. They have no nuclei.[27]

Our thinking about genes, development, and human identity is bound up in tired, incomplete, and frankly harmful metaphors. For instance, how well does the software metaphor work? Development relies on complex interactions among nuclear, intracellular, extracellular, and extraorganismic environments. Software for the most part relays inputs into outputs on a one-to-one basis. Complexity in software is an illusion generated by scale. Besides leading us to new ways to treat cystic fibrosis and male pattern baldness, one of the greatest boons from the human genome project could be the extinction of the myth of genetic determinism and genetic simplicity.

How can we encourage expensive research on potentially profound treatments? The genome project got this far with scant protection and no artificial shortages of data. Perhaps the hope of success, respect, glory, and the ability to save lives is incentive enough. The only option that seems to serve the public interest globally is to fund such research through state and multinational institutions. As a species, we should publicly pursue the goals of big medical science. We should not let the market guide our minds, lest we come up with a genetic treatment for baldness before breast cancer.

Science and Security

Commercial interests are not the only threat to open science. The ideology of the new security state has come into direct conflict with the ideology of science. The mysterious anthrax infections in Florida, Washington, D.C., and New York City in September 2001 led government officials to scrutinize the culture of science, which had a chilling effect. Scientists are demanding an honest assessment of the risks posited by open science and global collaboration. Four researchers at the State University of New York at Stony Brook released a paper explaining how a laboratory might synthesize functional samples of polio virus from the chemical precursors of polio's genetic material. The paper revealed that it is not necessary to start with living viruses to construct living viruses. The code, the chemicals that make up the viral coat, and the nucleotides that form the genetic code are all that's needed.[28]

The publication of the paper in the July 2002 issue of *Science* caused a flap. Soon U.S. government officials contacted editors of leading scientific journals to urge them to generate protocols for dealing with knowledge that might be used and abused by terrorists. After months of pressure and negotiation, more than twenty journals—including *Science, Nature,* and the *Proceedings of the National Academy of Science*—agreed to suppress the distribution of such articles. This debate continues in the scientific community.[29]

Regulating Math

Biologists are not the only researchers under pressure from commercial interests and national security concerns. If we boil down the conflicts between the electronic book industry and Dmitry Sklyarov, the recording industry and Edward Felten, the film industry and Emmanuel Goldstein, and the FBI and encryption researchers, they come to this: The U.S. government, at the behest of powerful media companies, is regulating mathematics.

Surely the traditions of free speech and free inquiry exempt basic mathematical research from the reach of the state. The fact is, math is everywhere. It is not just theory and conjecture, not just representations of reality in clear and simple language. Math, in the form of algorithms, influences everything and matters more than ever. To see the extent of this regulation, we must revisit the story of Dmitry Sklyarov, the second person to be charged with a criminal violation of the DMCA, which became law in 1998. The most onerous provision of the act makes it illegal to circumvent any protection measure that a copyright holder puts around digital files to regulate access. Most of the controversies about this law have been civil cases. Journalists for the hacker magazine *2600* have been fighting an injunction in federal court that prevents them from linking to Web sites that carry code for circumventing access control software on digital video discs. And in the spring of 2001 Princeton computer science professor Edward Felten was threatened with legal action when he announced plans to discuss his encryption research on digital watermarking of music files. In both civil cases the mere threat of DMCA enforcement stifled free speech. However, Sklyarov's employer, Elcomsoft, is a for-profit venture that sold his program as product. Felten and *2600* were just doing their jobs.

The law is clear. If you traffic in anticircumvention devices for profit, you have committed a federal crime that might cost you up to $500,000 and five years in prison. The FBI and federal attorneys have no choice. The problem obviously is the law itself. Congress was reckless in considering the DMCA in 1998. Some public interest advocates warned at the time that the law would stifle research, free expression, scholarship, teaching, and even commerce in new technologies yet to emerge. But Congress paid little mind to public interest concerns at the time. It was busy making the digital world safe for established software companies and movie studios.

What makes some programmers good guys and others bad guys? To be on the right side of the DMCA, you must be established. You must have already rolled out a product and hired some good lawyers. Introducing a new product that may disrupt an established one could be a crime. Instead of forcing Adobe to make a better product, the DMCA protects it with an inferior one. Still, Sklyarov and Elcomsoft did not commit a crime in the United States. Sklyarov wrote his program in Russia, which has no such law. His prosecution could look like a strike against non-American software firms, a nasty protectionist move, part of the effort to globalize American standards of corporate protection. Every effort to standardize and globalize meets strong resistance from those who suspect they might come out on the short end of it. It is hardly surprising that Russian software engineers take advantage of legal differences between nations to get a foothold in the global economy.

The Sklyarov episode is another chapter in the rocky story of the electronic book market. Publishers and distributors are betting that consumers are going to like reading electronic files on their computers. They hope we appreciate the price and convenience, the electronic indexing and bookmarking features, and the fact that one laptop can tote a shelf full of textbooks with no extra weight. Of course, there is no reason for believing people will actually like these products. Although there are regular frenzies of enthusiasm, all have fizzled. Still, publishers would love to instantly deliver their wares in a format that could be used by many people, would prevent unauthorized reading and copying, and would cost almost nothing for each additional copy. Paper books are expensive to make and distribute, and publishers never know how many to print. Often they are caught with an expensive surplus and take a big financial hit when hundreds or thousands of unsold copies are returned from stores.

The challenge of e-books is encryption. Selling one unencrypted biology e-textbook to one student in a three-hundred-person class might result in one sale and 299 unauthorized digital copies. Publishers want every reader to buy her own copy of a digital book. If they could tie a digital file to a particular computer (or perhaps make it disappear after a certain time or number of reads), they could recapture what they see as the "losses" of the real book economy—lending, rereading, photocopying, and reselling used books. But encryption comes at a high cost to users. When we buy encrypted content, we agree to limits on what we can do with it. We might sign away rights of fair use and first sale. While Russian and American programmers fight the encryption cold war, we can decide as consumers not to support the arms race at all. If I wanted to download and read the e-book *Enterprise Network Security Guidelines: Prevention and Response to Hacker Attacks,* I would pay an on-line retailer $132 and receive it as an unencrypted PDF file. I could read it on my desktop computer or share the file with my laptop and read it on the road. But if I want to download and read *The Unofficial Business Traveler's Pocket Guide: 165 Tips Even the Best Business Travelers May Not Know,* I would be out of luck. This file, which costs less than $10, only comes in the encrypted Adobe Acrobat eBook Reader format. It would be "tethered" to the computer on which I downloaded the file.

As we examine how distributed systems affect life in the twenty-first century, we see that encryption functions in a powerful and fascinating way. After September 11, calls went out from various congressional offices and some corners of the Bush administration that the United States should once again consider reining in the proliferation of strong encryption. There was widespread concern that terrorists had used encryption to evade surveillance by American intelligence officers. Alarming stories circulated of suspected terrorists using Internet-linked computers in public libraries.[30] It did not take more than a few weeks for U.S. leaders to review the encryption debates of the 1990s and realize the futility and counterproductiveness of such moves. Nonetheless, some very powerful people consider civilian access to strong encryption a threat to the state, children, and other living things. Meanwhile, in 1998 Congress criminalized the distribution of algorithms that decrypt access controls around digital content out of fear of the distributed nature of the Internet. Powerful content providers, including film, music, and parts of the commercial software industry, argued that allowing the distribution of such hacks would invite a flood of unauthorized copies of copyrighted material to flow freely around the world and undermine legitimate sales.[31]

While one arm of the U.S. Justice Department was calling for stronger regulation of encryption to deal with the threat of people passing coded messages over the Internet, another was prosecuting a Russian programmer for exposing flaws in weak encryption. The same contradiction plagues the world of finance. Authorities are concerned that strong encryption makes it harder to catch people laundering funds acquired through drug transactions or used to fund violence. On the other hand, electronic commercial and consumer banking could not work without strong encryption to ensure privacy and security. Encryption is just math. Any effort to restrict or enable the use or distribution of encryption is an effort to regulate mathematical algorithms. Any restriction on the discussion of such algorithms is a restriction on the progress of mathematical research. The battles over the use and abuse of strong encryption should make it clear that the answer to our problems is neither more tech or less tech, high tech or low tech. Technology only does what we make it do. It can be designed as a protocol or as a control; it can be used to liberate or debilitate. And both protocols and controls can have unintended consequences.

Between Heston and Heidegger

Encryption doesn't kill people—people kill people. The slogan "guns don't kill people—people kill people" expresses what philosophers of technology call the "neutrality thesis." Philosopher Andrew Light calls this the "Hestonian" position because actor Charlton Heston, when he was president of the National Rifle Association, relied on this argument to defend the relatively free flow of firearms and ammunition through American society. As Light concedes, "there is nothing about the distribution of guns in and of itself that increases the propensity toward violence." Mid-twentieth-century thinkers such as Jacques Ellul and Martin Heidegger, reflecting on the horrific killing power of dehumanizing technocratic regimes, effectively countered the neutrality thesis by generating what Light dubs the "substantive" view of technology: the very culture of technology (what

Ellul called "technique") shapes the way humans and societies think and behave. Those who take the "substantive" position on technologies conclude that certain channels of behavior, enabled in large part by the technologies themselves and often unexamined or debated by society at large, are inherently harmful to the process of achieving meaningful happiness, either individually or collectively.[32]

Between technological neutrality and technological demonization, between Heston and Heidegger, lies a method of examining technologies in context, in history, and in the best available light. Technologies do embody ideologies. Some reflect those ideologies more than others. Some technologies—such as the general-purpose computer or the distributed network—are complex enough that they might embody multiple or even conflicting ideologies. They are not merely extensions of our appendages and organs. They are built and used for specific purposes, but not always in the ways the builders expect or predict. In fact, encryption can kill people; it can also save lives. Consider one of the most interesting distributed information systems in the world today: FreeNet.

FreeNet is a fully distributed peer-to-peer network that ensures the privacy of its users through the use of strong encryption. It is almost perfectly ungovernable, except to the extent that it is "governed" by its design and protocols. It is so well distributed that it is impossible for any element of the FreeNet community to hijack the system and bend it one way or another. FreeNet users trade text, music, video, and other files with complete confidence that no outside power—Internet service provider, cable company, university, or the secret police of the People's Republic of China—can monitor the transactions. For political dissidents in Burma or China, using encryption and FreeNet to share information could mean the difference between life or death, and certainly the difference between freedom and torture. FreeNet shows us that strong encryption can be a tool for enabling or extending communication in illiberal contexts, just as it can be used to limit communication in other contexts. It's not exactly neutral technology, but it's not hitched to a single ideology either. Encryption is at the hinge of the struggle between information anarchy and information oligarchy.[33]

FreeNet is an open communal programming project. It's licensed like free software, with a requirement that all improvements to the open source code be revealed and shared. The free software movement is the best example of the progressive, efficient, and democratic effects of distributed systems. As Yochai Benkler explains in his paper "Coase's Penguin," what he calls distributed peer production can make better, faster, more flexible tools for communication and cultural production. Benkler argues for more systematic study of the cultural and economic power of highly energized distributed systems. As Benkler notes in his paper, the culture of free software is the culture of science. The operative protocols of science are publication, peer review, and revision.[34]

Like the free software movement, traditional science is under threat from powerful interests that consider openness a threat to the dynamics of investment, control, and profit. A study published in the *Journal of the American Medical Association* revealed that 47 percent of geneticists surveyed about their information-sharing habits reported having at least one request denied over a three-year period. Nearly a third of these impeded researchers reported they could not pursue their research questions without the denied information. Increasingly, the report found, commercial interests have been chilling genetics researchers from sharing important information.[35]

Science and math are founded on open communication and global citizenship. Scientific political structures are loose enough to take the best parts of radical democracy and refine them with a strong adherence to protocol. The culture of science built new global infrastructures through which we share music, poetry, data, and rumors. We should endeavor to protect and celebrate our successes while being honest about the risks inherent in openness and freedom.

CHAPTER 10

The Nation-State Versus Networks

Just yesterday, it seems, influential thinkers were imagining a world in which the nation-state would wither, and many decisions that affect everyday life would be shifted up to multilateral institutions or down to individual market actors. Technologies were to play a leading part in that change, linking cosmopolitan citizens and transnational markets in a way that would enable crude governance, cultural creolization, and efficient commercial transactions. Human beings were on the verge of finding new and exciting ways of relating to one another. Arbitrary barriers of ethnicity and geography would wither. Through technology, we would master the dynamics of, and therefore control, our "cultural evolution."[1]

This vision was informed by a synergy of soft anarchism, market libertarianism, and techno-fundamentalism. It assumed that the state would eventually slough away. Meanwhile, we would have to push and prod it to relinquish centralized control over daily matters. The tautology went: This sort of radical globalization was going to happen anyway—the technology determines it. So its advocates urged personal and policy choices that would guarantee it. If those left out of the global, technocratic, educated elite were to suffer a bit, that would be the price of cultural evolution. We would wire their villages and gently inform them of the impending changes. In practice, of course, the instruments of this particular form of globalization did not serve the softly anarchistic vision of a decentralized species acting in concert. Like Lenin, who deferred proletarian rule, we would have to centralize authority in corporate boardrooms and multilateral confederacies until all the villages were wired. The commercialized World Wide Web—America Online—would be the operative medium of cultural evolution. FreeNet would not.

Now we see that the nation-state is not going anywhere. Ethnicity and geography still matter, and we may even be experiencing a "cultural devolution." If anything, the nation-state has capitalized on the mania of "globalization" and "information" to reinforce its powers and jurisdictions. We had a moment of techno-globo-utopian idealism in the 1990s, but clearly the nation-state is back and the dominant form of globalization is oligarchic, not anarchistic. So the most pronounced forms of opposition to that dominant model are understandably informed by anarchism.[2]

That's not to say that the nation-state is what it once was or that it will behave in the same ways in the future. The pressures on state sovereignty, identity, and security are significant. The plain fact that people, currencies, culture, and information are more portable than ever has increased the anxieties that nation-states endure concerning identity and security. Pressures come from inside and outside, as immigrant groups retain some interest in the politics and culture of their homeland and expatriate communities dispersed around the globe, willingly funding and enabling new challenges to state security and integrity. Different pressures on sovereignty come from above and below—from oligarchic multilateral governing institutions and from teeming anarchistic mobs of techno-libertarians and disgruntled rebels. These can be described as the "Washington consensus" and a strange synergy between the "California ideology" and the "Zapatista swarm."

Soft Oligarchy: The Washington Consensus

The Washington consensus was a form of market fundamentalism complicated by bad faith. Although its advocates claimed to champion "free trade" and "open markets," there was nothing free and nothing open about the Washington consensus. It was more Washingtonian than consensual, existing mainly among major institutions in Washington, D.C., and representing the vested interests of developed nations. While intending to empower market forces, it depended on coercion by institutions that resemble superstates with no direct democratic accountability. In practice, powerful multilateral institutions such as the World Trade Organization, the International Monetary Fund, and the World Bank determine important policies of many nation-states. Clearly the multilateral institutions that enforced the Washington consensus serve the interests of a handful of rich, powerful states in North America and western Europe.[3]

Born in the 1980s and matured in the 1990s, the Washington consensus was a set of principles that many states agreed on through a series of multilateral negotiations. Advocates of the consensus claimed that free trade among nation-states would allow each to shift resources to what each produces most efficiently. No individual state should rig the global price game through trade quotas, tariffs, subsidies, or price supports. If Japan produced superior automobiles at a lower unit price than the United States, then over time and over a level playing field Japan would displace Detroit. The United States would have to shift its investments in human capital and heavy machinery into some other realm of production such as software or soybeans. The rules of arbitration and review that push nation-states to abandon industrial protection policies are enforced by the World Trade Organization (WTO), which was established in 1995 after several years of negotiations.

The WTO is a forum for resolving trade disputes. If Japan accuses the United States of artificially supporting its steel industry through import ceilings or tariffs, it appeals to the WTO for review. If the United States accuses Japan of dumping cheap steel on the world market to undermine prices and drive competitors out of business, it does the same. Before 1995, trade disputes were often handled bilaterally (through trade wars or contentious and complicated diplomacy) or through the terms of the General Agreement on Tariffs and Trade, the agreement that preceded the WTO. But the WTO is considered a much better forum for resolution because its membership is broader and its governing structure clearer. The WTO facilitates "liberalization" of trade by providing a forum for cooperation among states, reducing the transaction costs of challenges to a state's policies, and generating and maintaining norms of behavior.[4]

"Free trade" is only one part of the Washington consensus. It demands that developing nations tighten their budgets so that they do not contribute to currency inflation, which can cause inefficiencies in the flow of goods and capital while allowing debtors to repay loans with cheaper money. In exchange for the opportunity to take advantage of loan programs and currency reforms engineered by institutions like the World Bank and the International Monetary Fund, developing nations have been pressured to reduce the role that the central state plays in the economic lives of their citizens. This means that no one is accountable to citizens who need government aid. Elected leaders may feel bound to adhere to the directives of the World Bank and International Monetary Fund, which undermines their abilities to meet the needs of their citizens, or they feel that the Bank and the Fund are likely to bail them out of spendthrift endeavors, so they act irresponsibly.[5]

The Washington consensus was an oligarchic solution to one of the problems created by the rise of democracy as the preferred method of governance around the world: Democratically elected leaders serve their constituents. It's hard for a legislature to ignore the pleas of a textile industry that employs thousands of voters and is crying out for relief from the price pressures of cheap imports. If legislatures around the world actually served the stated needs of their established industries, they would lock advantages into their methods of production. They would offer tax breaks and extra levels of legal protection to powerful segments of the economy. New ideas might die for lack of investment. The government might end up subsidizing a sick, decrepit industry that has no business competing with sleeker, smarter competitors abroad (or, more likely, competitors that pay their workers less and adhere to lower environmental standards). To avoid the problems created by democracy, states that adhere to the Washington consensus give up a lot of sovereignty for ideological stability.

If, for instance, the Brazilian legislature decided to develop an indigenous passenger airplane industry, it would consider a whole slew of industrial policies. Lacking a national rail system and quality highways in much of the country, Brazil relies on air travel to link its citizens and markets. The legislature would consider such policies as massive development of domestic airfields, tariffs to limit the importation of passenger planes from the two dominant global producers—Boeing from the United States and Airbus (a consortium funded by France, Germany, and England)—and direct subsidies for research and development. This would lessen national competition (and thus gestation) but increase global competition over time (eventually three instead of two major airplane makers). These efforts would also generate a fierce backlash from Europe and the United States, which would complain to the WTO of unfair trade policies closing lucrative Brazilian markets to their planes. The United States would also exert pressure through the IMF, which might threaten to tighten credit or repayment terms on Brazil's massive debt. Brilliant, creative Brazilian aeronautical engineers might endeavor to work for Boeing in Seattle or Chicago instead of São Paulo or Rio de Janeiro. The system is rigged in favor of established industries in rich nations. Thus the logic of the Washington consensus offers no true level playing field, no truly open markets, and no freedom to innovate and compete. The logic ignores important historical factors in the development of economies like France and the United States, which subsidized their domestic aeronautical industries for decades. Before the WTO and the Washington consensus, there was no way to stop them. Boeing is the product of massive government spending and high levels of protection. Without similar strategies at their disposal, developing nations don't stand a chance to compete.[6]

In another perverse, undemocratic consequence of the Washington consensus, a legislature may act irresponsibly (as the U.S. Congress did in 2001) by supporting high steel tariffs that it knew would be struck down by the WTO. A government may satiate the desires of interest groups while avoiding the risks and results of actually implementing protective policies. The real policymaking power gets shifted upstairs to the WTO, where unelected arbiters decide which environmental, cultural, and labor policies are retained and which are discarded. In this way, the Washington consensus pressures states to relinquish sovereignty by asserting the powers of global, multilateral governing structures that have no electoral accountability and very little transparency. Citizens of nations around the world feel the effects of these institutions every day. Yet they rarely get to voice their needs before them.[7]

Techno-Libertarianism: The California Ideology

From about 1981 through 2000, the Washington consensus offered a new political order: a weakened, less relevant nation-state in the twenty-first century with a stronger supernational structure of oligarchic control. But a revolution was brewing on the left coast of the United States that encouraged the passive erosion of state influence on markets and people's lives. Participants thought it was a revolution, declared it a revolution, and acted as if it were a revolution. Although it turned out to be less revolutionary than many hoped, its ideological influence was undeniable. It never achieved its libertarian dreams, yet it enabled the spread of cyberanarchy. Political economists Richard Barbrook and Christopher May call it the "California ideology," but it might more properly be called the "Northern California ideology."[8]

The California ideology predicted that new communicative technologies linking consumers directly to producers (sans middlemen) and allowing consumers more and faster information for decisionmaking would radically alter global capitalism. Transaction costs would fall. Consumers would demand better quality and service at lower prices, and the smartest firms would supply it. Workers would no longer be tied to offices and plants. Managers would slough away as corporate hierarchies collapsed. Employees would find greater satisfaction working contract to contract for a variety of firms on individual projects instead of latching their fortunes and reputations to just one firm. Firms would "outsource" much of their work, from printing to data storage, to shipping, to research, to accounting. At every level—consumers, labor, management, and the firm itself—everyone would be a free agent. Firms that worked better with their minds than their muscles would win. Work would be flexible, workers would be free, and social needs would be served through private ventures that capitalize on quick application of knowledge and networks of experts. More than withering because private firms would serve consumers (who used to be called citizens) better, the nation-state would be actively dismantled because its interventions perverted the flows of information that would fuel this revolution in the first place.[9]

One of the founding gurus of the California ideology was management consultant Peter Drucker. Through such well-balanced, sober works as *The Age of Discontinuity*, Drucker predicted that "knowledge techniques" would build this emerging "knowledge society." Although he never called for a radical erosion of the state, Drucker valorized the brainy over the brawny by preaching management flexibility and adaptability, and advocated a shift of resources to research and development. Drucker's cool modesty made his work more valuable to managers than pundits, and it has retained its value for more than fifty years.

The work of Drucker's less-than-cool counterpart, George Gilder, has not held up so well. Gilder was responsible for some of the great excesses of the California ideology. Gilder was the prophet of profit, the necromancer of network effects, and the oracle of Oracle. Through his "must-read" investment newsletter, *Gilder Technology Report,* he sent bad money after bad as he pumped up one dot-com stock after another through the 1990s. He claimed to discover "laws" guiding the new economy and thus guaranteed the success of the firms he pushed as "inevitable." Financial observers noted the "Gilder effect": Share prices rose and fell based on what Gilder wrote about them. In his 2000 book, *Telecosm,* published just before the stock market left many of his chosen companies discarded on the trading floor, Gilder predicted that the massive surplus of fiber-optic bandwidth for electronic communication—supplied partly by Gilder-inspired investments in the then-valuable company Lucent Technology—would soon create a revolutionary "end of scarcity" of communication. This would be a historic change, Gilder claimed, surpassing even the invention of the personal computer (the death of which he proclaimed in the first line of the book). Chock-full of philoso-babble, Gilder's books and reports fueled the spread of the California ideology by emphasizing that doubters need not apply. They would be discarded and ignored by the rush of history and the triumph of technology. You couldn't fight the revolution, Gilder claimed. Resistance was futile. You might as well invest in it and enjoy the ride to riches.[10]

The California ideology found its most bombastic voice in *Wired* magazine executive editor Kevin Kelly, who saw in "network" dynamics a grand unifying theory, an almost spiritual catharsis. Inspired by the flowering of the Internet, Kelly envisioned human society moving toward a beehive status, workers moving and shifting along loose networks of communication, ignoring limits and barriers, rendering state control superfluous. Like so many others, he saw networks everywhere, and they explained everything he liked. Kelly made simple, declarative, predictive statements like "encryption always wins because it follows the logic of the Net." For Kelly, the logic of the Net was permanent and stable, and states would bow before it once they got hip to the power of network dynamics and their self-regulating ways. In the preface to his 1998 book, *New Rules for the New Economy,* Kelly sounded very sure of himself: "No one can escape the transforming fire of machines," he pronounced. "Steadily driving the gyrating cycles of cool technogadgets and gotta-haves is an emerging new economic order." Kelly described this "new economy" as something more than a tweak or bubble in our familiar analog economy of soybeans, softball bats, and shopping malls. "This new economy represents a tectonic upheaval in our commonwealth, a far more turbulent reordering than mere digital hardware has produced."

Kelly went on to make a claim not too far from one I make in this book: The world of atoms would soon be governed by the rules of bits. "The logic of the network will spread from its base in silicon chips to infiltrate steel, plywood, chemical dyes, and potato chips. . . . consider oil." The imposition of more technologically advanced drilling tools, he predicted, would radically reduce the cost of extraction and thus the price of crude oil. But Kelly took no account of the cost-raising effects of shipping, processing, risk management, cartels (global and local), and cleaning up the pollution caused by oil. He seemed unfazed by the possibility that atoms, in the forms of bombs and bullets, might affect the market for oil more than bits of information ever could. The difference between Kelly's observation and mine was that he celebrated this change and its potentially revolutionary and liberating efficiencies, while I am bearish on the applicability of such ideological constructs and consumed by the high human costs involved. Kelly failed to consider that oligarchic powers of the real world can bite back, anarchistic reactions can undermine commerce, and atoms can trump bits like scissors cut paper.[11]

The proliferation of the California ideology, a self-fulfilling prophecy reinforced by temporary and unjustifiable stock price surges, contributed to what economist Edward Luttwak calls "turbo-capitalism." Luttwak wrote that by the 1990s, political and economic leaders in the United States and Europe had religiously accepted some unexamined beliefs—state influence on economic decisions encourages inefficiency, misappropriation of resources, and general complacency. The patron saint of this ideology was the grandest Californian of all, President Ronald Reagan. These fundamentalist assumptions had yielded a massive rollback in the role of individual states in their essential roles as curbs on the excesses of capitalism, a role that had facilitated the unparalleled economic success of the United States, Canada, France, Germany, Japan, and the United Kingdom after World War II, and of Taiwan, South Korea, and Singapore in the 1980s. Turbo-capitalism justified, among other things, massive deregulation of the electricity and natural gas markets in many parts of the United States in the 1990s and thus led to the rise and ultimate collapse of the paradigm of turbo-capitalist enterprises: Enron.[12]

Although it was based in Houston, Texas, Enron preyed on California. It was the corporate embodiment of the California ideology. It was a "virtual" company that attracted billions of investment dollars and the deep loyalties of political leaders of both major American political parties. It ran on faith, gumption, fraud, and—contrary to its free market fundamentalist propaganda—expensive political connections. Enron produced nothing; it claimed it made money by trading futures and options on ethereal goods and services like energy and fiber-optic bandwidth. It was Gilderesque in its techno-fundamentalism. It was Kellyesque in its religious belief in networks. It was Reaganesque in its optimism. While offering sunny reports about its financial standing and earnings, Enron was hiding debt in bogus offshore "shell" companies, capitalized only on the promise of future shares of Enron. Back in the real world, Enron was creating much misery and using its political connections with the Republican Party to cover its sins. Throughout the summer of 2000, while electricity prices in California shot up beyond all reason and San Francisco Giants fans growled at the audacious Enron sign in left field at Pac Bell Stadium, Enron executives (and later Vice President Dick Cheney) claimed that supply could not meet demand because California had not deregulated enough. A year later regulators and economists revealed that Enron and other companies had participated in a massive price-fixing scheme to gouge Californians. Just months after the California electricity scandal, the company collapsed under the weight of its own doctored financial statements. In December 2001 Enron filed for bankruptcy protection. Thousands of loyal employees lost their jobs and their retirements funds, which were heavily invested in worthless Enron stock. Faith, ideology and corruption were not enough to keep Enron going.[13]

Someone who believes that networking and digital technology will radically alter the economies of the world, unleashing a flurry of joyous creativity and liberty, may well leap to the conclusion that the institutions which serve and curb those economies—nation-states—will change as well. After all, the modern nation-state evolved in tandem with opportunistic global trade and later modern corporate capitalism. If communication systems were going global, and corporations were getting virtual, then the state must be ready for some serious changes as well. "A new phe-nomenon, the virtual state, has arisen," wrote political scientist Richard Rosecrance in 1999. "Lodging production abroad enables this new kind of state to specialize in higher-valued, intangible goods: products of the mind." The chief engines for this change would be—of course—technol-ogy, global flows of information, and direct overseas investments, as the state itself diversified, globalized, and shrank. The model states in the twenty-first century would look more like Singapore and less like Japan. And the rise of the virtual state presaged the irrelevance of armed conflict (ah, to remember life in 1999). The virtual state echoed the new virtual corporation, and one facilitated the other. The California ideology, it should be noted, almost always (except in the case of Enron) generated celebrations and predictions of great things ahead for Californians, or at least upper-middle-class, overeducated, cosmopolitan, libertarian-leaning Californians. Conveniently, those who wrote the treatises of this ideology were upper-middle-class, overeducated, cosmopolitan, libertarian-leaning Californians.[14]

Caffeinated Anarchy

In some ways growing directly out of new communicative technologies fostered by the California ideology and in other ways brewing up from dis-gruntled subalterns in developing nations, anarchy burst into relevance and importance in 1999. It filled the streets of Seattle and shut down a round of negotiations at a meeting of the World Trade Organization. But even as it rose as an ideology to be reckoned with, anarchy remained widely misunderstood and mischaracterized. Activists from all corners of the Earth had been searching for ways to challenge the Washington con-sensus. They finally found inspiration in the 1994–1995 Zapatista uprising in Chiapas, Mexico.[15]

Since the Spanish conquest in the seventeenth century, Mexico has been ruled by a light-skinned, Spanish-speaking elite. Two anticolonial revolu-tions failed to empower indigenous ethnic groups. As the ruling party in Mexico's one-party republic pushed its policies closer to the Washington consensus, rural indigenous people felt the brunt of the changes. Thanks to the North American Free Trade Agreement (NAFTA), American-made maize, grown with heavy subsidies from the U.S. government, flooded Mexican markets, displacing those who had grown maize for centuries. Store-bought (and even American-made) tortillas filled store shelves, leaving women who made fresh tortillas for a living without jobs. Young women from rural areas were recruited to work long hours in dangerous factories along the Texas border, often tied to their jobs for years to pay the debt incurred from their recruitment and transfer.

Such social disruption inspired a variety of responses among the poor in Mexico. The most successful and inspirational came from Chiapas. The Zapatistas' chief ideological spokesman, Subcomandante Marcos, did little real commanding. Instead, he outlined the Zapatista agenda: respect for indigenous civil rights, the establishment of real democracy in Mexico, and an abrogation of NAFTA. The Zapatistas adopted anarchistic tools and tactics such as organizational transparency, global networks of communication, and passive resistance but were not fully anarchistic. Because they represented the poor, illiterate, and communicatively stranded, an educated elite handled many of the decisions within the loose organization. Still, the movement struggled to limit the powers of any one leader or faction and remain true to its democratizing mission. While the Mexican government tried to portray the rebels as violent thugs willing to resort to atrocities to split up the republic, the Zapatistas quickly responded through decentralized communicative channels, fax machines, short-wave radio, and newsletters. Striking ironic militant poses with toy weapons, they soon convinced international observers that the Zapatistas were largely nonviolent (in contrast to the brutally violent government response) and that their appeal was aimed at Mexican civil society at large. Unlike violent revolutionary groups such as the Palestinian Liberation Organization, the Irish Republican Army, or the Shining Path in Peru, the Zapatistas requested that international human rights and relief agencies come to Chiapas to monitor their activities. They denied Marxist influence, ties to communist governments, or a utopian vision for Chiapas, Mexico, or the world. They only sought social justice and real democracy in Mexico. Unable to defeat the Zapatistas with propaganda, the Mexican government sent 12,000 troops after them. Hundreds died as both sides resorted to violence. Once the Zapatista regulars withdrew to the forests, where Mexican soldiers were at a disadvantage, the overt violence ceased. Soon Mexican officials rounded up peasants to interrogate and torture. As news of torture spread, Mexican officials prevented human rights groups and journalists from visiting the area. These tactics reminded many Mexicans of the brutal crackdowns on student protesters in Mexico City in 1968. If 1968 represented the apex of the Mexican "cultural revolution," the 1994–1995 Chiapas uprising was Mexico's Tiananmen Square and inspired hope for a "velvet revolution." Such a revolution remains elusive. But the Mexican government soon learned that it could not afford to evoke global revulsion and be recognized as a modern, democratic state. Mexican civil society experienced a revival, and in 1999 Mexico held its first legitimate multiparty elections since the rise of the dictatorial and corrupt Institutional Revolutionary Party (PRI) in the 1930s. It also tolerated a futile student uprising in Mexico City that shut down that city's leading public university for more than a year.[16]

Using the slogan "The Revolution Will Be Digitized," activists all over the globe took direct inspiration from the Zapatistas. Anti–Washington consensus parties in Venezuela and Brazil won elections in the early 1990s. Meanwhile, Mexican voters, many of whom have benefited from increased trade with the United States, elected a conservative president who once worked for Coca-Cola and lived in the United States. Now that it has a president chosen by a fair plebiscite, Mexico is the second-most democratic nation-state in North America after Canada. European anarchists and activists helped Zapatistas organize the First International Encounter for Humanity and Against Neoliberalism in Chiapas in 1996. Through subsequent meetings in 1997 and 1998, the movement spread to include several important trade unions in Europe and Canada.

These activists sought true and complete globalization. The partial, rigged globalization promulgated by the Washington consensus tied workers to one place. It encouraged the movement of money, resources, and goods but did not allow the free flow of people and ideas (unless these ideas were encased in Hollywood films and music, and then only under strict market, legal, and technological controls). A free flow of people and ideas, activists argued, would affect authoritarian states by allowing them to sense deep threats grumbling up from their subjects; multinational corporations could not exploit wage differences effectively enough to undermine unions. Many diverse groups forged a movement with the appearance of incoherence, which in fact reflected a coherent diversity of concerns and methods. "We declare," the founding document of the movement read, "that we will make a collective network of all our particular struggles and resistances, an intercontinental network of resistance against neoliberalism, an intercontinental network of resistance for humanity." As sociologist and anarchist David Graeber wrote, this new global anarchism is not only pro-globalization—in the sense that it hopes to erode borders and allow people to seek fulfillment wherever and however they might imagine—it is a rare major social and ideological movement that has spread from the South to the North, from developing to developed nation-states. In their effort to define a bond with humanity over nation as a first principle, these activists echoed Diogenic cynicism.[17]

Diogenes found an ideal playground in Seattle. Anarchists, environmentalists, labor union members, farmworkers, and other critics of the Washington consensus shut down a meeting of the World Trade Organization there in the fall of 1999. The home of Microsoft, Boeing, and Starbucks Coffee was also a node of global communication and the flow of tourists and workers. Its economic success in the 1990s made Seattle the ideal showcase for the Washington consensus. But its proximity to Native American communities and old-growth forests made it a symbol of all that the Washington consensus threatened. The technologies that the WTO celebrated in Seattle—intercontinental air travel, large quantities of cheaply grown caffeine, and unmediated global digital communication—undermined the institutions that supplied them. The ruling institutions of the world were shocked and completely unprepared, since they had read anti–Washington consensus activists as fragmented, unsophisticated, and lacking widespread public support. Most accounts falsely labeled the protest movements as antiglobalist when they actually supported globalization. They were falsely labeled violent when they were most definitely antiviolent. As in Chiapas, the government perpetrated the violence once the activists' tactics overwhelmed their ability to make sense of the situation.

For the most part, the Seattle activists practiced "direct democracy." The loosely affiliated groups were themselves composed of loosely affiliated members. They ruled themselves through protocols. When a member proposed an "action," she invited participation and criticism. After deliberation and debate, members who opposed the revised proposal could opt out of the action. In response to extreme proposals that violated the core principles of the group, members could propose a veto, and the group then considered the validity of the concern and decided whether to act. Such loose consensus can degenerate into organizational paralysis, but the more urgent the issue and the more reasonable the action, the more effective these organizations can be. Once these movements shifted from the conference and seminar rooms—and chat rooms and Web pages—to the streets of Seattle, they were much more flexible, impressive, and effective than anyone in power (or in universities) could have predicted. The Seattle activists were mostly, in Graeber's term, "small 'a' anarchists," as opposed to more overtly ideological "Anarchists." Like the Zapatistas, they dabbled in anarchistic tactics and methods without overtly endorsing a stateless world vision.[18]

Efforts since 1999 to replicate the triumphs of Seattle have been frustrated by events outside the activists' control. Impressive protests in Quebec in the summer of 2001, intended to stop progress on a Western Hemisphere trade treaty on the model of NAFTA, failed to stop the talks but sent a message. In contrast, protests in New York City intended to disrupt the World Economic Forum meeting in early 2002 were largely unimpressive and ineffective. Between these two events, of course, the World Trade Center fell, and citizens and states around the world shifted their immediate concerns from globalization to security. In Genoa in July 2001, Italian police gunned down a young man named Carlo Giuliani who was protesting the meeting of the G8, the eight most powerful economies in the world. Amid 80,000 protesters who were calling for cancellation of Third World debt, a police vehicle ran through crowds of mostly peaceful demonstrators to strike back against a handful of violent protesters. Many were chased and beaten. In Genoa, the idealized vision of "anarchists with a small 'a'" evaporated as a more extreme, uncompromising faction reverted to violence against Italian security forces and world leaders, some lobbing Molotov cocktails over barricades. These violent anarchists seemed out of place in the global movement inspired by the Zapatistas. Yet their actions—and blow back from the conservative Italian government— have become part of the governing mythology of the battle over globalization. The protesters basked in glory after Seattle, and right-wing Italian authorities had no interest in seeming overwhelmed, surprised, or incompetent, as Seattle police had. This combination of hubris and militant defensiveness had fatal consequences for progressive forces in general and Carlo Giuliani particularly. As activist Nathan Newman explains, "There was, I think, a somewhat un-strategic overconfidence that developed among protesters post-Seattle. The Seattle cops were unprepared and played into the propaganda goals of the protesters. As Philadelphia and now Genoa showed, the cops are no longer unprepared and are developing both the repressive technology and propaganda to crush the Black Bloc-style protesters and the rest of the movement if we don't develop some new strategies to control the escalation of violence."[19]

By 2003, these three ideological challenges to the power of the nation-state seemed stalled if not dead. Their influence altered the expectations of capitalists, anticapitalists, and leaders around the world. But their long-term effects were not easy to measure by the end of 2001. Under the leadership of two dissimilar nation-states, the United States of America and the People's Republic of China, the twenty-first century opened with a clear call to think nationally first, and globally only if a clear, direct payoff to the nation-state could be achieved. The ideologies and networks that seemed to threaten the nation-state in the 1980s and 1990s faced challenges far greater than the nation-state ever did.

CHAPTER 11

The Empire Strikes Back

By 2003 Enron had exposed the soft underbelly of the California ideology, Genoa had stunned the Zapatista swarm, and Washington had unraveled the Washington consensus. The administration of George W. Bush publicly endorsed the principles of free and open markets but did nothing to enable or execute them.[1] In fact, the administration took many steps to undermine the Washington consensus and reestablish American sovereignty and protectionism. It established high steel import restrictions. It reestablished strong subsidies for American agricultural production. It eroded the power of multilateral institutions generally by abandoning agreements on the International Criminal Court and the Kyoto Protocols that aimed to reduce greenhouse gas emissions. And it declined to pursue a second, more specific United Nations Security Council resolution as a trigger for war in Iraq, opting to invade without international authorization. While evading multilateral commitments, the Bush administration pursued bilateral trade treaties with countries such as Chile, Morocco, and Singapore.

Instead of peaceful, open borders among the three major countries of North America, in 2002 and 2003 the lines to cross into Mexico and Canada grew longer and longer, and immigrants faced increased scrutiny and the real threat of detention. In the rush to profit from the devastation in Iraq in 2003, the United States held closed auctions for construction contracts and limited the bidders to American firms—in direct violation of WTO rules. In all of these actions, the Bush administration avoided the mitigating power of the "international community." Some began to wonder if there were such a thing as an "international community" or "global civil society," a concept that had seemed matter-of-fact just a few years before.

From Civil Society to Civil Disobedience

In the fall of 2002, *Foreign Policy* magazine asked a series of experts to define the international community. All of the responses questioned whether individual nation-states can organize and execute policies through multilateral institutions. In the long, tall shadow of belligerent U.S. unilateralism and its "cowboy" swagger, the writers doubted whether the "international community" exists in a stable and definable form. The term itself seems to represent a claim to global legitimacy for a policy aim or goal, yet reflects nothing more than a loose consensus among world leaders from developed nations. For some respondents, such as the anarchist Noam Chomsky, U.S. domination of the agenda for such organizations undermines the myth that "international community" reflects the will or desires of "international civil society."[2] But as U.N. Secretary-General Kofi Annan saw the situation, the rules and procedures of international organizations had a record of expressing the will of a "civil society" and influencing world events in often positive ways. Yet Annan allowed that the institutions of global governance are "hardly more than embryonic." In setting out the need for a well-defined "international community," Annan summarized what might be called the "global condition" or "global moment," as people in disparate parts of the world see how their daily lives are changed by forces beyond their control. "We are all influenced by the same tides of political, social, and technological change," Annan wrote. "Pollution, organized crime, and the proliferation of deadly weapons likewise show little regard for the niceties of borders; they are problems without passports and, as such, our common enemy. We are connected, wired, interdependent."

Annan's assessment of the global condition is significant on two levels. First, it cites several problems that threaten the future of Homo sapiens and defy the ability of nation-states to contain. Consequently nation-states have organized into institutions to share burdens, exchange information, and combine resources. Second, Annan makes it clear that we are connected by things more essential than radio waves, wires and cables, and global financial flows. First and foremost, we share four oceans and one atmosphere. Annan is pleading, without an overt declaration, for reasonable, dependable, authoritative, republican forms of organization and implementation. He cites the International Criminal Court as the best example of a humane and humanitarian institution that provides a desirable alternative to inconsistent, hypocritical, or ad hoc reactions to crimes against humanity. The United Nations, of course, was set up as such an institution.[3]

The problem with Annan's sober exhortation for both the existence and the necessity of an "international community" is that most of the influential institutions that claim to speak and act for it actually represent elite interests. The United Nations, with the broadest claim to global representation, is far from a truly representative body. It boasts 191 member states, yet the permanent seats on the Security Council enjoy undue influence that (with the exception of the People's Republic of China) fails to reflect the proportion of human beings represented. As long as the tiny United Kingdom is a permanent member of the Security Council while Indonesia, India, Nigeria, Mexico, and Brazil are not, the United Nations will not truly represent the "international community," let alone "international civil society." The World Bank and International Monetary Fund, which exercise wide-ranging influence over the lives of billions of people in developing nations, clearly work for the interests of the developed nations. The policies and decisions of these two institutions come straight from the U.S. government. The World Trade Organization has perhaps the worst record of global representation. As Annan concedes, "the international community is a work in progress."[4]

This problem of global legitimacy has not gone unexamined. As Arjun Appadurai has noted, "the international community is neither international nor a community." It does not exist except as a moral idea invoked by politicians and pundits. It is not an institution in itself, nor does it have a process or procedure through which it can express itself. And it does not act as a community. Yet it is a compelling moral idea, Appadurai argues, because it reflects wishes born in the democratic revolutions of eighteen-century Europe. Because the formerly colonized parts of the world are still learning to assert their interests on the world stage, the established states have been able to rig the system in the favor of their own interests. "Thus, as a social and political reality, the international community does not inspire any real sense of ownership among the poorer 80 percent of the world's population," Appadurai writes. "And even among the upper 20 percent, it remains a network for a relatively small group of politicians, bureaucrats, and interventionist opinion makers." The operational "international community" is, in other words, a global power elite.

While the "international community" speaks from the position of stable nation-states (and stable, limited interests within those nation-states), citizens of nations at all stages of development and enduring a variety of states of unfreedom forge powerful links through new media, old postal systems, and the very oceans and atmosphere that we all necessarily share. Groups of mobile or extranational communities—Kurds, Romani, Chinese emigrants, educated technologists, Nigerian expatriate dissidents, North American farmworkers—carry multiple loyalties and multiple forms of cultural and economic citizenship.

These dispersed peoples, who link themselves via elaborate communications networks and define themselves via "portable" versions of their heritages and histories, constitute the world's great hope for some general appreciation or recognition of the global human interdependence that Annan described. A family in Karachi, Pakistan, with members living well in Manchester, Montreal, and Milwaukee, is less likely to view the world as a simplistic and dangerous "clash of civilizations." It knows firsthand that one economy influences all others, and war and fear in one part of the world soon infect another. As Appadurai explains, the "mental geography" of an increasingly important segment of the world population does not match the official geography of the United Nations. "In this sense," he writes, "these communities mimic the global marketplace, which is now strikingly beyond the regulative capabilities of most nation states. . . . They are symptoms of the impossibility of constructing new global organizations on an international conceptual foundation."[5]

Here Appadurai evokes one of the most powerful indicators that anarchy matters. In December 1999 the city of Seattle was overrun by protesters of various stripes, all of them upset about how the WTO was going about its business. The Seattle protesters were more than soldiers for radical democracy. They gave us a glimpse of what an authentic "international community" might look like. Instead of being based on some hard-to-demonstrate will of "civil society," these alliances of grassroots activist groups represent "uncivil society." They did more than speak for the disenfranchised, those excluded from "civil society." They *were* the disenfranchised. Their unhierarchical methods of organization and deliberation were, in some ways, the methods of more "civil" than traditional, elite "civil society." In other ways, less so. Heady from their success in Seattle yet tempered by less impressive efforts to disrupt global elites meeting in Quebec, Davos, Prague, Washington, D.C., and New York City, more than 50,000 activists gathered in the Brazilian city of Porto Alegre in February 2002 as the World Social Forum. They met to reconnect, coordinate, share experiences, and forge manifestos and alliances. Naomi Klein, addressing activists at Porto Alegre, declared, "The alternative to a world without possibility," Klein proclaimed, "is not civil society—but civil disobedience."[6]

If civility is no longer an option, states and their subjects will resort to uncivil means of exerting their will. In terms of the regulation of culture and information, this has driven the policies of states as different as the People's Republic of China and the United States of America in similar directions. Both recognize threats from inside (immigrant or dissident communities) and outside (foreign or diaspora communities) as communicative systems worthy of surveillance and disruption. States of all kinds have a compelling interest in stifling flows of information that could contribute to disruptive violence (like terrorism), the sexual exploitation of children (like child pornography and slavery), and organized crime (like money laundering, drug trafficking, and prostitution). Authoritarian states also declare a compelling state interest in preserving a static political culture. They do their best to stifle diversity, dissent, and corrupting influences (ethnic cultural expression, pro-democratic content, and pornography, however the state chooses to define it). In the post-Tiananmen, post-Seattle, post-Napster, post-9/11 age, the state has reasserted itself by trying to manage its information ecosystems. Because information ecosystems flow into each other, this inevitably entails trying to manage other people's ecosystems as well.[7]

Networks of Terror

States are constantly identifying nefarious "networks" at the heart of every perceived threat. Some networks are internal but have foreign sources and influences. Others are entirely alien. The rhetorical value of alleging a "network" at the heart of a threat to security or identity is clear: It's impossible to tell when a war against a network is over because it can't be seen. A network can be dispersed, distributed, encrypted, and ubiquitous. As the 2002 national security strategy of the United States of America asserts:

Defending our Nation against its enemies is the first and fundamental commitment of the Federal Government. Today, that task has changed dramatically. Enemies in the past needed great armies and great industrial capabilities to endanger America. Now, shadowy networks of individuals can bring great chaos and suffering to our shores for less than it costs to purchase a single tank. Terrorists are organized to penetrate open societies and to turn the power of modern technologies against us.[8]

Such a description yields a broad, almost unlimited set of prescriptions, all of which substantially increase the surveillance and police powers of the state. In a war against an enemy that cannot be measured, cannot even be seen, there is no way to limit the set of tasks that the emergency seems to justify. And there is no way to declare the emergency over. If the network can't be seen in the first place, how can we be sure it's gone? Here the security strategy is clear:

To defeat this threat we must make use of every tool in our arsenal—military power, better homeland defenses, law enforcement, intelligence, and vigorous efforts to cut off terrorist financing. The war against terrorists of global reach is a global enterprise of uncertain duration. America will help nations that need our assistance in combating terror. And America will hold to account nations that are compromised by terror, including those who harbor terrorists—because the allies of terror are the enemies of civilization. The United States and countries cooperating with us must not allow the terrorists to develop new home bases. Together, we will seek to deny them sanctuary at every turn.[9]

Technology plays two roles in this scenario. "Our" technology must have wide berth. We must invest in it without limits, regardless of the actual threat or the efficiency and effectiveness of the technological defenses we build. But if "they" get hold of "our" technology, we are in big trouble, for "they" are ungovernable.

The gravest danger our Nation faces lies at the crossroads of radicalism and technology. Our enemies have openly declared that they are seeking weapons of mass destruction, and evidence indicates that they are doing so with determination. The United States will not allow these efforts to succeed. We will build defenses against ballistic missiles and other means of delivery. We will cooperate with other nations to deny, contain, and curtail our enemies' efforts to acquire dangerous technologies. And, as a matter of common sense and self-defense, America will act against such emerging threats before they are fully formed. We cannot defend America and our friends by hoping for the best. So we must be prepared to defeat our enemies' plans, using the best intelligence and proceeding with deliberation. History will judge harshly those who saw this coming danger but failed to act. In the new world we have entered, the only path to peace and security is the path of action.[10]

The enemy is ungovernable because it exists as a nefarious network. Thinking about networks in many areas of life is important and revealing and is useful when appropriately applied. But many popularizers of "network theory" see networks in everything from social groups to economic principles to terrorists. Such thinking is becoming dangerous and illiberal. In early 2001 the Rand Corporation released a book called *Networks and Netwars*, arguing the enemies of peace and stability will be networks, not nation-states. Since 9/11, the Pentagon and White House have been taking this book way too seriously and wrongly applying it to the threats facing the United States. You can see the seeds of Total Information Awareness and the USA Patriot Act in the prescriptions offered by *Networks and Netwars*.[11]

It didn't take long for one moral panic to inspire another. Articles quickly popped up comparing terrorist organizations like al Qaeda to a distributed system like Napster. They argued, much as *Networks and Netwars* did, that if al Qaeda were a distributed network, then the United States should be poised to "hack the network." And that's just what it set out to do. A school of thought inspired by *Netwars* offered the following syllogism to policymakers: (1) Terrorists operate as a distributed network. (2) The only way to hack a distributed network is to corrupt the data and monitor the end points. (3) Therefore, to fight terrorists we should come up with distributed strategies to fight a distributed network.

These theories have inspired some specific policy proposals:

- Corrupt the data: The Pentagon wanted to distribute misinformation and propaganda through news media. This plan was allegedly put on hold by the Department of Defense after word got out. Of course, the very idea of disinformation casts doubt on the Pentagon's assertion that it canceled the program.

- Monitor the end points: The Justice Department considered instigating TIPS (terrorism information and prevention system), described in some of the literature as "open source intelligence," which would deputize private citizens to spy on one another.[12]
- Mine data: The Defense Department proposed collecting private financial data from hundreds of companies and monitoring citizens through such programs as Total Information Awareness.

But what if al Qaeda is not a distributed system? Wouldn't this "hacking" merely expand the surveillance power of the state while leaving Osama bin Laden's plans undisturbed? Several articles since September 11, 2001, have posited that al Qaeda is a network comprising independent cells that make it unquashable. Therefore the U.S. government must abandon its policy of bombing and invading and instead find a more clever, more distributed way of attacking it. While I applaud the call for better understanding of distributed systems, I think associating al Qaeda with networks is both alarmist and frankly wrong.

To the best of our knowledge, al Qaeda is centralized, top-down, and heavily dependent on the cult of personality of Osama bin Laden. Al Qaeda (which literally means "the base," a concept absent from distributed systems) distributes human capital, weapons, and perhaps funding around the globe. But orders flow one way. Resources are distributed so the organization as a whole lacks vulnerability. Information, however, is centralized, regulated, and revealed tactically. Few members in any given "sleeper cell" know plans, share intelligence, or have the option of including or excluding themselves from an ordered action. Even Peter Bergen, the journalist who has described the inner workings of al Qaeda, has fallen in love with high-tech metaphors. Calling it "al Qaeda 2.0," Bergen argued in a *New York Times* op-ed piece in November 2002 that when the United States bombed al Qaeda out of Afghanistan, it "morphed into something at once less centralized, more widely spread, and more virtual than its previous incarnation." Yet none of the evidence (al Qaeda had, since September 11, 2001, launched a handful of attacks that involved conventional explosives and easy-to-commandeer vehicles) supported Bergen's assertion that the group was fundamentally transformed. Its communicative dexterity was certainly weakened and its leadership is harder to find, but there is no reason to believe that al Qaeda is not still a top-down organization with hierarchically arranged operatives in many places around the globe. "The base" may have moved from Afghanistan to Pakistan, but it did not transform itself into something "virtual." Al Qaeda remains actual.[13]

In contrast to the small band of radical thugs who form al Qaeda, the global community of Islam, the Ummah, is a distributed system. Islam has had no ruling authority, no caliph, since 1924. It has open membership sans credentials. Membership is based on adhering to protocols, believing, praying, adhering to rules and guidelines, and spreading the word. Ummah supports a vast diversity of practices and interpretations within standard and universal protocols. No one licenses authority in Islam. Elements of Ummah reward certain scholars and leaders. No one appoints or certifies expertise in Ummah. As a result, as we are frequently reminded, Taliban leader Mullah Omar is not a real mullah, yet he persists in calling himself one.[14]

If I were to draw technological metaphors, I would call al Qaeda a distributed denial-of-service attack on Ummah and Islam, an attempt to hijack the network. But we don't need such characterizations to understand how a small, driven faction can struggle to represent, if not control, a larger, more diverse array of actors. Al Qaeda and its partner, the Muslim Brotherhood, is one highly centralized branch of Ummah, which has no mechanism to purge such a malignancy. To many outsiders looking at Islam after the attacks of 2001, the hijackers had taken over the system. But a sophisticated analysis of the diversity and dialogue within Islam tells a very different story. Most Muslims never embrace the stern extremism and hatred that fuels al Qaeda, Wahhabism, and other forms of Islamist political radicalism. Since 2001, mainstream elements of Islam have clearly set themselves apart from the more dangerous elements of the faith. Openness and debate within Islam have kept al Qaeda and its clones marginal.[15]

How should we fight al Qaeda? Bombing and infantry attacks seemed to work through spring of 2003, when the Bush administration foolishly shifted military and intelligence resources from Afghanistan to Iraq, which offered no threat to world security. The *Netwars*-inspired strategy yielded the arrest of hundreds of Arab Americans and immigrants, but none were charged with a terrorist-related crime and a thick wall of secrecy fell over administrative trial proceedings that determined their fate. Specious theory generated brutal policy, and controls replaced protocols.

What's a State to Do?

The rise of distributed information systems certainly complicates, and perhaps undermines, the sense of stability in a nation-state, which is itself an information system. Among its various roles, the nation-state regulates flows of information; a nation-state can even be defined as a particular method of information management. Along with enthusiasm for torture and concentration camps, a characteristic that distinguishes a totalitarian state from a democratic republic is how it restricts the flows of certain pieces of information and the technologies that carry them. An authoritarian state is the source of "official" information, and all other information is suspect.[16]

In its fight against distributed information systems it deems dangerous, the United States has been considering ways to exploit the wealth of information about everyday people gathered by private databases.[17] Through the ill-fated (now decentralized and renamed) Total Information Awareness program and other similar, less publicized efforts, the U.S. government hopes to cross-reference habits and tendencies in a way that reveals the intentions of dangerous people. This effort requires constant government access to huge databases of credit card purchases, hotel reservations, airline systems, bank records, tax records, voter registration records, property deed records, automobile registration records, and rental records. The investigators use complicated indexing algorithms to sort through this pile of data looking for suspicious trends and patterns. If a particular person left an electronic trail of questionable activities that (if taken independently indicate nothing alarming) might warn of impending danger when considered collectively, law enforcement agencies would act.

For instance, if I, an American citizen with a South Asian name and an olive complexion, were to travel from Miami International Airport to Cairo via Milan (as I did in January 2002), I would not arouse suspicion. But if my data trail indicated that I had paid cash for my tickets (which I did not), and that I had also traveled to Atlanta and stayed in a hotel near the Centers for Disease Control in the previous week (which I had not), and bought several tons of fertilizer (which I have never done), the government might decide to watch me more closely. At first glance, this method of law enforcement seems alarmingly intrusive. For centuries, law enforcement has used the starkly ineffective method of profiling suspects based on ethnicity and class, and the data-mining practice, comparatively, can be applauded for being ethnically neutral. We would all be subject to Total Information Awareness, and that's precisely what alarms citizens most. The innocent—especially those from ethnicities not accustomed to state scrutiny—don't want to be watched any more than the guilty do.

In fact U.S. citizens are data mined and profiled constantly. Companies compile data and sell it to other companies that build huge data sets. From these data sets, to which many other companies subscribe, marketers can estimate or guess a person's income, clothing preferences, next car, and dietary restrictions. Data mining presents multiple problems for law enforcement that go beyond civil libertarian discomfort. For one thing, privately compiled databases are often wrong. To serve commercial purposes, they only need to furnish mostly accurate (rather than completely accurate) information to justify the investments in them. Inappropriate direct mail never hurt anyone. For example, some friends of mine shopped regularly at the Food Lion in Maryland. They used a discount card that created a record in the Food Lion database that tracked their brand preferences. After noting that my friends seemed to be buying their meat somewhere other than Food Lion, the store barraged them with coupons. However, the database had no way of knowing that my friends are vegetarian. Such a failure in data analysis cost Food Lion a little money but caused no harm. The system overall probably made up for whatever loss the chain suffers from its vegetarian shoppers. In other words, data mining is good enough for commerce and marketing. But law enforcement and security demand higher standards of accuracy. The difference between Food Lion and the U.S. Department of Justice is that the government can imprison you.

In the uproar over Total Information Awareness, one particularly strong criticism of data mining came from Gilman Louie, chief executive of Q-Tel, a firm founded by the Central Intelligence Agency in 1999. Speaking at a computer industry meeting in March 2003, Louie warned that if the federal government used automatically updated data-mining systems to create a "watch list," it might be impossible for an innocent person to remove herself from the list, especially in cases in which people of the same age share a name. The innocent person would be removed from flights, denied security clearances, stricken from jury lists, and watched constantly. Data mining is blunt and impersonal. False positives are frighteningly common. Even if the rate of false positives were a remarkably low 1 in 100,000 adults, the American population would yield at least 1,600 false positives. If the false positive rate is a more realistic 1 in 10,000 adults, then 16,000 Americans would be falsely identified. It took only ten terrorists to fly planes into the Pentagon and World Trade Center. To find these ten (and there is no reason to believe that such a data-mining system in 2001 would have), 16,000 innocent Americans would have been snared in the nets of federal investigators. False negatives—perhaps more importantly—are common as well. A terrorist who guesses the criteria that the government uses to track data can rig the game. She can use false documents, stolen credit cards, and off-shore venders. Total Information Awareness and its less alarmingly named cousin programs would make us less secure by giving us a false sense of security. The bad dudes can and will evade the net. The rest of us will get tangled once in a while.[18]

Louie suggests a more subtle data analysis technique that relies on actual investigative work instead of random distributed behaviors. Once a law enforcement agency reasonably suspects someone of illicit behavior, officials can use databases to track associations with other known threats, such as terrorist organizations, organized crime, or dangerous individuals. Then they can paint a broad portrait of behaviors and patterns—addresses, neighborhoods, bank account access, credit card use, travel—that the suspect and the known danger share. Even this system would harm innocents who happen to know bad people through casual contact. But it might serve as the beginning of a real, intensive, human-driven investigation. Innocent Canadians might still get deported to be tortured in Syria because they had the wrong roommates years ago. Such horrendous mistakes are likely to multiply if the U.S. government installs a stupid data-mining system.[19]

We would expect to see oppressive, brutal states (e.g., Burma, Saudi Arabia, and the People's Republic of China) monitor or impose technological restrictions on communicative technologies at their citizens' disposal. But even relatively liberal nation-states such as the United States and the United Kingdom are taking steps to block or monitor legitimate flows of information without the usual concerns for due process or free speech. The sovereignty of information policy is up for grabs these days. Because many of these initiatives are emerging through multinational organizations and corporations, citizens have no forum for debate or appeal. And policies justified in an atmosphere of panic have no chance of being fairly or rigorously debated.

The Locust Man

Liu Baiqiang understands the power of distributed information. While serving a prison sentence for theft in 1989, Liu Biaqiang wrote pro-democracy messages on tiny scraps of paper and released them into the air on the legs of locusts. According to a report by the Supreme People's Court, the messages included slogans such as "Long Live Freedom" and "Deng Xiaoping Should Step Down."

Because of these messages, in June 1989 Liu Biaqiang was sentenced to eight years' imprisonment for "counterrevolutionary incitement" and "propaganda." He was held in Shaoguan prison in Guangdong province and was not due to be released until the end of 2003. Liu was one of many prisoners convicted after unfair trials in the aftermath of the 1989 massacre whose cases were never reviewed.[20] Liu was released in the fall of 2001 after international human rights organizations championed and publicized his cause.[21] His story remains important because of his courage and creativity, and because he demonstrated a principle of communication that is beautiful in its simplicity: Powerful forces can't always stop small things from jumping long distances.

As the People's Republic of China tries to quiet democratic activists and stifle religious practice, it is experiencing an acute tension between the nation of "greater China" and the state of the People's Republic of China. As Ian Buruma describes this tension, the diaspora composed of ethnic Han peoples who form communities from India to Singapore to Australia, Guam, Vancouver, Chicago, and London have a vision of Chineseness and the greatness of China. These communities maintain their cultural ties through mail, movies, music, and the Internet. Many of the members of this diaspora live in democratic states, and some under authoritarian states. Still others, if Taiwan and Hong Kong are included in "greater China," have experienced slippery versions of both. The margins of "greater China" are more religious, more democratic, more liberal, more idealistic, more optimistic, and more connected than the 1.3 billion who live under the constant surveillance of the People's Republic of China.[22]

As the example of greater China shows, it's important to distinguish a nation from a state. Matthew Arnold wrote that the state is "the nation in its collective and corporate capacity." A nation is a collection of people with a bond that links them over time and space. The members of a nation need not fill a particular space at a particular time. They need only imagine that they share a history, a condition, or an identity. A state, on the other hand, is a function—a collection of administrative powers and institutions. It may claim legitimacy through democracy or through security, or by speaking for the presumed interests of the nation that it may attempt to map. But, as in the case of the nation of greater China overlapping the state of the People's Republic of China (or Kurdish communities crossing the borders of two states—Turkey and Iraq), a nation need not conform to the map or the power of the state.[23]

The experiences of the People's Republic of China are central to this exploration because its efforts to regulate information flows have been the most overt, daring, and controversial in the world. Chinese government efforts to regulate access to the Internet through licensed Internet service providers, to stifle the use of proxy servers, and to quash the use of encryption by private citizens have all been made in partnership with major American commercial software firms—the same firms that Richard Clarke wants to consult to construct his new and improved Internet for "us," to protect "us" from "them."[24]

FalunGong.net

In late November 2001, the People's Republic of China conducted a coordinated sweep of 94,000 Wang Ba, or Internet cafés, and shut down 17,000 of them. Many Wang Ba had failed to install the required filtering and blocking software—controls, not protocols—that would block access to sites sponsored by the distributed, decentralized, open information system known as Falun Gong. As a result of this action, almost ten years after the student uprising filled Tiananmen Square and filled the hearts of democrats everywhere with hope for the future, the Chinese government faced another demonstration. This time, tanks and guns were not as effective. Hundreds of followers of a fast-growing religion gathered in Tiananmen Square to meditate. Their organized silence frightened China's leaders, who had no idea that the sect had such a strong following in that officially atheistic country.[25]

The Chinese government reacted fiercely. It arrested hundreds of Falun Gong supporters and set about "deprogramming" them, often with jail time and torture. But it had no way of rounding up leaders: Falun Gong did not seem to have any. No single person had "called" or "ordered" the demonstration in the square. Yet the members seemed united, calm, ordered, and of one mind. Perhaps jealous of the devotion it generated, the Chinese government turned to methods that would disrupt the "network." Falun Gong acts like a distributed network. It has a spiritual leader who lives in New Jersey but no pope, no central authority. It grows by being open and adaptable. It threatens the Chinese government by being uncontrollable, and members find each other through the Internet if they live outside the People's Republic. They use pay telephones if they live inside.[26]

Some religions operate according to well-defined hierarchies. The Roman Catholic Church is the best example. Other faiths have hierarchies embedded in distributed systems. For instance, the oppressive authority that Shiite mullahs exert over the Iranian state and Iranian society contrast starkly with the way Islam works in the lives of most Muslims around the world. When such centralized power structures emerge, they stifle diversity of thought, action, and belief among believers. Judaism and Islam share much more than roots in the Middle East and direct descent from Abraham (or Ibrahim). They are both largely distributed, open systems. To be or become Jewish or Muslim, you need only adhere to protocols. The precise number and nature of those protocols are constantly debated within both religions.

In response to the distributed threats that democratic activists and religious practitioners seem to present to the stability of the Chinese state, authorities tried to regulate the infoscape of the larger nation. Clearly many of these disruptive influences came from abroad—from the Chinese diaspora, from greater China. The People's Republic of China would have to reengineer its interface with the outside world—a monumental task. Reengineering would involve registration and surveillance at Internet cafés. It would require that all Internet users on the mainland get access through approved state-run portals. These portals would demand constant staff monitoring to block proliferation of sites offering pro-democratic or pro-religious content. China has been trying to build and extend its electronic commerce potential while simultaneously limiting what its people read and how they communicate. The open, protocol-based, radically democratic, distributed network of networks that we call the Internet does not seem ideal for these purposes. But it's cheap, easy, popular, and already existing. So the Chinese government decided it would roll out Internet access on its own terms, which meant controlling to the best of its ability the last mile of the network. In the meantime, the People's Republic of China would try to build its own proprietary network, its own vast Intranet, over which it would have substantial control.[27]

Hacktivism

As soon as the oligarchs in charge of the Chinese Communist Party started building the great firewall of China, hackers started punching holes in it. A group of hackers—some members of the nation of greater China, others passionate advocates of democracy or anarchy—started producing technologies that allowed dissidents to communicate beyond the gaze of the Chinese state and access materials the state deemed contraband. The same tools have been used by dissidents from authoritarian and totalitarian states as diverse as Burma, Yemen, Cuba, and Laos. The strategies to protect dissidents start with concealing the location and identity of the user. Thus encrypted communication is central to this effort. E-mail encryption must be strong enough to frustrate state security forces yet transparent enough for dissidents to use and adapt rather easily. "Hacktivists" have maintained "proxy servers," through which Chinese users may access forbidden sites by appearing to tap into unapproved or unknown servers.[28]

The most ambitious effort to aid dissidents attempts to simultaneously hide the identity and location of the user and allow unfettered access to contraband. It's a classic peer-to-peer system: FreeNet. Founder Ian Clarke published an academic paper outlining its principles. Basically the system allows users to piece together content held anonymously in fragments by a wide array of servers around the world. It takes content away from the World Wide Web, which is too public for the most seditious content, and distributes it among hundreds of volunteer hard drives. The system privileges security, privacy, and redundancy. FreeNet protocols resolve the addresses where the content lies and reassemble it for the user, while strong encryption keeps the IP number of each user secret. FreeNet is hard to use, but it's impossible to stop. It's the perfect Internet: radical, anarchistic, and potentially influential and important peer-to-peer electronic networks. Napster was a goof, but FreeNet is revolutionary.[29]

Failed States

But revolution is not enough. Anarchy is not sufficient for building a life or a society. When open communication and organization are forbidden, clandestine communication is necessary and justified. Although we may advocate anarchist tactics in places like Burma and China, we must beware the destructive, tyrannical power of majorities whipped up by "loose ties" of nationalism, ethnicity, or religion. The Chinese Communist Party fears anarchy in any of its forms, by any of its definitions. Its public pronouncements, its criticism of what it calls "western" democracy, and its effort to cleanse its 1.3 billion people of distracting and disruptive thoughts speak directly to its conviction that a little anarchy can lead to a lot of chaos, a lot of disorder, and a lot of trouble. Since the fall of the Communist bloc, literal anarchy in the crudest sense—the absence of effective government— has raised its head a number of times. We can see what the Chinese government fears in Rwanda, where the majority tyrannized the minority and slaughtered millions, and in the Ivory Coast. Journalist Robert Kaplan posited in his book *The Coming Anarchy* that in the early years of the twenty-first century, many parts of the world would start to resemble the Ivory Coast. Local ethnic and religious tensions would erupt, panics would ensue, and no form of imported or imposed "civil society" could stem the bloody chaos. Every person would have to fight for his or her life. Social ties would break like straw. Personal and political alliances and the trust that supported them would crumble. Kaplan predicted the steady erosion of state control in much of the developing world.[30]

An astute critic of Kaplan's "methodology," political scientist Jennifer Widner, has responsibly cataloged the measures and metrics of "failed states" and successful states in sub-Saharan Africa. "The conditions that prevail in the countries that have disintegrated do not constitute a Hobbesian war of all against all," Widner wrote, directly contradicting Kaplan's predictions. "The ferocity of conflict is apt to be much lower where people act as individuals and make no effort to cooperate." The real danger in collapsed states comes from the state's lack of authority, Widner asserts, often undermining faith in itself via corruption, arbitrary execution of policies and laws, and general incompetence—and from the "weak ties" that bind a group and transform it into a violent mob. These mobs do not emerge from an absence of "social capital" but from toxic levels of it, forged by imagined bonds of ethnicity, kinship, or religion. Networks of civic engagement actually contribute to a breakdown in larger civic structures. "Conflict becomes more devastating when the possibility for coordinated action exists," Widner remarks. "The collapse of states in Africa occurs when people still feel some social obligations to one another but where the boundaries of the moral community parallel ethnic divisions." The problems start when individuals in a crowd deny the legitimacy of the authorities, and this sense of delegitimacy spreads through unmediated, uncensorable channels. "The milling about that one observes in crowds just before hostilities break out reflects efforts to try to determine whether there is sentiment for a collective denial of the rights of civil authorities— a condition under which the risk of punishment for being 'the only' or 'the first' to break the rules disappears."[31]

This sort of movement, as noted in Chapter 1, is "half-baked" anarchy that only endorses the elimination of authority and governance. It offers nothing but passion and violence. "It is not necessarily the case that there must be a new norm everyone shares in order to trigger this kind of reaction," Widner writes. "All that must take place is the withdrawal of rights of control from established authority. Thus, it is possible for this kind of collective action to take place with no evident vision of a goal or a better way to run the government." Widner writes that half-baked anarchy clearly emerged in Rwanda in the bloody mid-1990s. "The exclusivist militants and some of the Hutu governors were distributed throughout the country. This decentralization facilitated 'milling' and their ability to gather information about whether others were prepared to overthrow existing norms."[32]

"Smart mobs" go bad when communicative anarchy succumbs to majoritarian passions. Half-baked anarchistic systems that do not include habits and structures of reasoned deliberation are vulnerable to exploitation by wily, manipulative demagogues. The bloodshed in Rwanda and the Ivory Coast is exponentially graver than battles over global copyright piracy, but the parallel dynamics are worth considering. Some combination of tight controls, lack of moral legitimacy, breakdown in social norms, and a means to spread corrosive information contributed to an anarchistic crisis, to "uncivil disobedience." When norms crumble, people feel justified in breaking laws and the state feels compelled to break heads.[33]

CHAPTER 12

Conclusion

The Heartbreak of Oligarchy and Anarchy

The heart of my argument in this book is a call for modesty and patience. I have avoided making predictions. I have declared no wars. I have refrained from declaring a new historical epoch. I merely plead that the variety of human experience is so vast that we should not try to guide it from above. We also should not shrink from the challenge of discussing and debating—and thus clumsily guiding—how we will govern, share, and organize information systems in the new century. This book is about the politics of culture and information. It's not about censorship per se or the invisible structures of media, politics, and public expression, what Walter Lippmann described as the "manufacturing of consent." My concerns are the availability and accessibility of the substance of expression and thus the possibility of public discussion and creativity.[1]

This book was supposed to be about entertainment—the battle over control of digital music, text, and video—an extension of my first book, *Copyrights and Copywrongs: The Rise of Intellectual Property and How It Threatens Creativity.* But as I researched this new project, the world shifted beneath my feet. Borders melted. Buildings collapsed. Thousands of my neighbors died, crushed by metal and concrete. Seemingly invincible corporations withered and crumbled. Dreams evaporated. And my government betrayed many of its core values. By 2004 the future of Metallica's fortune didn't matter much. My concerns moved to the regulation and control of all sorts of information, much of it cultural, much of it political. "Information warfare" became an important concept by the spring of 2002, as the U.S. Justice Department refused to reveal information on hundreds of people it had detained for months without issuing charges or explanations, and the Defense Department declared it would spread lies publicly to aid its antiterrorist operations.[2]

Multilateral organization and alliances seem to matter less. Yet the only sovereignty that seems worth defending is American sovereignty. According to my government, previously public information is now contraband. Too much truth is dangerous, and more lies make us safer. Meanwhile, the U.S. government has instituted policies that strengthen its ability to gather and interpret information about its citizens. It will know much more about us and we will know much less about it. And we will suspect our neighbors of complicity in nefarious activities and crimes. Information flows became more important than ever as the U.S. government declared it a major front in its foreign, domestic, and military policy. Because I am an educated, privileged American charged with studying flows of information and how they affect people's lives, I felt it was my duty to address these issues. This book grew in response to the demands of the times. I could not get up in the morning to write any other kind of book.

The transition from entertainment politics to information politics was less complicated than it may seem. After all, much of politics is about entertainment and much of entertainment is about politics. In *Copyrights and Copywrongs* I concluded that the United States corrupted its copyright system by privileging corporate interests to the detriment of the public interest. Reckless expansions and extensions rendered copyright an instrument of censorship, just as it was in the seventeenth century, when the state did the censoring. By the end of the twentieth century, corporations did the censoring and the state merely facilitated the practice. Most importantly, the United States set about reengineering the very tools it claims are the source of the problem: anarchistic technologies of digitization, networking, and encryption. In addition to playing an important regulatory role in culture, copyright serves as a canary in a coal mine. The strategies that are emerging in copyright battles resemble those in more important battles over democracy and human dignity.[3]

The role of the nation-state is in flux, and its future is up for grabs. Those who are struggling to control flows of information—for very good reasons of commerce, security, and stability—include legislators, judges, cabinet officers, leaders of multilateral regulatory institutions, university presidents, corporate executives, lobbyists, and generals. Those who are trying to liberate information include students, educators, librarians, computer programmers, civil libertarians, religious leaders, artists, consumers, political activists, and dissidents living under oppressive regimes. But the whole story is not that simple. Most people have complicated relationships with the politics of information, favoring freedom by default but acknowledging the need for regulations under certain conditions.

As I explored these stories, arguments, and issues, I found that the systems of information regulation were getting messier and more complicated. The exploding popularity of distributed peer-to-peer networks seemed to undermine centuries of assumptions about the costs, speed, and ease of information transfer. Hackers were flaunting their powers to disrupt regulatory mechanisms, battling to keep their anarchic space free. Simultaneously, nation-states, from the United States of America to France to India to the People's Republic of China, began to rein in the troublesome aspects of the Internet, proposing restrictive and potentially effective technologies on top of the old favorites like laws, prisons, fear, and torture. Meanwhile, the same corporations that corrupted the American copyright system were expanding their regulatory powers globally and strengthening their enforcement of oppressive new laws in the United States. But no one was winning the war. New devices kept appearing that were supposed to fix the problems caused by the previous devices. Every time one party made a move, the other two reacted. Having just emerged from the hangover of the last global arms race, we now faced a new, less deadly, less expensive technological race. Those of us who stood outside these forces looked on with alarm and bemusement. The music was free, but the stakes were much higher than the price of music.

Deadly Synergy

The urge to break heads, to do the bidding of oligarchy by any means necessary is intimately linked to specters of anarchy. The urge toward anarchy depends on oligarchic abuses. Each creates the conditions that allow the other to thrive. The question for us in the twenty-first century should not be choosing anarchy or oligarchy but constructing and maintaining systems that discourage both. Anarchy is a reaction, not a vision or solution that can produce the best society and the best human future. Anarchy offers itself as a reasonable set of tactics only because democracy and liberalism seem so fragile, so easily corrupted or evaded, so readily claimed yet so rarely realized. The shift of decisionmaking power from the local to the global elite has invited an anarchistic blow back. The rapid, inexpensive connections of our global age have put the tools of anarchy into the hands of many, who have helped spread and shape the ideology. Because our tools helped teach us how to be anarchists, the oligarchs have tried to dull them.

By the early twenty-first century the tools of anarchy—pager, mobile phone, and networked computer—were widely available. The ideology of anarchy offered the only obvious alternative to the oligarchic vision of corporate and state control of everyday life. Perhaps appropriately, it was these same state and corporate oligarchs who made many of the technological tools of anarchy available and affordable. If we were upset about the policies of Monsanto, McDonald's, Merck, Miramax, or Microsoft, all we could do was opt out of commercial transactions or smash some windows. There was no other recourse because there was no forum for discourse. Since much of everyday life was ruled from afar, the only answer to a lack of accountability seemed to be inarticulate protest, through mass demonstrations in the streets of Seattle or massive uncivil disobedience in the form of peer-to-peer downloading. The world at the dawn of the twenty-first century has many needs, among them a cure for malaria and the second coming of Bob Marley. In the meantime, we could use a reinvigoration of two ideologies left out of the battle between oligarchy and anarchy: rich cultural democracy and healthy civic republicanism.

For Cultural Democracy

In the past twenty-five to thirty years, the U.S. government decided to champion the interests of a handful of companies, and most cultural and technological policy has shifted in their favor. Policy about who gets to own and run networks, who gets to own and run radio stations, how long copyright protection lasts, what forms copyright protections take are dictated by lobbyists and only rarely rise to the level of public debate.

Both democracy and creative culture work best when raw materials are cheap and easy and easily distributed. Any cultural development that has made a difference in the world—reggae, blues, needlepoint—is really about communities sharing, moving ideas between and among people, revising, playing with theme and variation, and ultimately forging consensus about what is good and what should stay around. This is how culture grows. These behaviors are older than cassette tapes or motion pictures and have been amplified and extended by the powers of digital technology and networking. Quantitatively we are in a new situation, although qualitatively we are not. We're actually behaving as we always have.

Culture is worthless if you keep it in your house. In this sense, the proliferation of shared culture—ostensibly free material—is simply the electronic simulation of what people have been doing in towns and villages and neighborhoods and garages and high schools around the world for centuries. This sort of creative circle—the drum circle or the blues-singing circle—is a vivid image of these creative communities. They form any place artists gather, any place musicians jam for the fun of it. This is an important part of being human. It's the essence of being cultural.

I don't think we want a world that offers no incentive to produce *Star Wars* or *Casablanca* (although we might imagine a world without Jar Jar Binks). We might imagine a world in which someone could write a sequel to *Casablanca* and not be laughed at. It's this notion of working from our shared cultural phenomena to build new and special things. That's why we need low barriers of entry to creative processes and cheap access to cultural materials—through electronic networks, friends sharing material, public libraries, universities, schools, temples, mosques, and churches. These are all institutions built for sharing.

We need to ask bigger questions about these processes. Shouldn't our policy priorities be diversity and easy access to culture and information? Shouldn't we demand local input on matters of culture and politics? Shouldn't we allow churches to set up small radio stations to serve their constituents? Shouldn't we allow political activist groups of all flavors to do the same? Shouldn't we strive for lower levels of regulation of culture and information? Shouldn't we chop up our electromagnetic spectrum in a way that maximizes the number and variety of voices that can flow over it? Shouldn't we encourage new democracies to build information regulation systems that reflect local needs instead of Disney's? We need to examine these issues ecologically and then generate a set of principles and policies we can live with.

As cultures build themselves and proliferate, they pretty much follow anarchists' description of the ideal political state. Anarchy is not the ideal political state—far from it. But anarchists are onto something descriptively. Culture builds itself without leaders. Culture proliferates itself through consensus and revision. Culture works best when there is minimal authority and guidance. We must declare a desire for global cultural democracy. In 1974 sociologist Herbert Gans declared support for cultural democracy when he saw critics and snobs as the chief impediments to a full appreciation of the variety and levels of cultural production and the values of sharing. Now, almost thirty years later, critics and snobs are no longer serious threats to cultural democracy. Rather, it is states and corporations, deploying laws and technologies, that threaten cultural democracy.[4]

For Civic Republicanism

Accepting the anarchistic nature of culture means looking at the systems in which oligarchy is imposing itself and generating panic about information anarchy. The horror stories may be justified; for instance, some anarchistic information systems allow bad people do bad things. No one can defend information anarchy when it makes child pornography easily available and widely distributed. Yet radical software systems like FreeNet make it just as easy for people to share child pornography as the works of Thomas Paine and Dr. Dre. Freenet is built for irresponsibility: No one can remove anything from it, and thus no one is responsible for it. Those of us who celebrate the freedom of these new information systems tend to ignore the very bad things that can go on through these systems. We should never celebrate irresponsibility.

The real question is, What methods do we use to attack bad things such as child pornography, white supremacy, terrorism? Do we build new machines that block these flows of information? Is that good in the long term, and, just as importantly, is that harmful to those of us who want to use those systems for good? These real problems are complex and have deep roots in history, and confronting them effectively is going to take decades or centuries. However, we have been trying to confront them technologically and shallowly.

We reach for easy answers. When someone suggests a quick fix to a problem, we should ask two questions: Would such a system be effective against the real problems? And would the harm that comes from that sort of intervention outweigh the benefits? The second question is really hard to answer. So let me ask them both a different way: If this technological intervention is effective—and that's a big if—is there a less intrusive way to achieve the same result? If surveillance of everybody might deter a handful of terrorist acts, is there a way to imagine more targeted surveillance? Could we lower the cultural cost of security? Could the surveillance be based on hard work and real investigations, instead of being comprehensive and intrusive? Could surveillance be based on trust between the government and its citizens, so that they feel invested in the public good? The best way to stop any illegal act, terrorist or otherwise, is to make sure that terrorists do not have support in society in general. We do this by making sure life is good and secure so that the people around those ne'er-do-wells have some sort of investment in or loyalty to the larger community. There are complicated, hard, messy ways to attack crime, terrorism, and general misbehavior. They are expensive and imperfect. But they are ultimately more effective and more likely to engender trust.

Nurturing public trust is the essence of a political ideology—civic republicanism—that has been left to wither in recent decades. Republicanism has its roots in the thought of Aristotle in Athens and Cicero in Rome. Revived in the age of the city-state by Machiavelli, revised with the rise of the nation-state by James Madison, and refigured by such twentieth-century thinkers as Hannah Arendt, republicanism seeks to mitigate the excesses and weaknesses of individualistic liberalism and thick communitarianism. Recognizing that liberalism (and, at its extreme, libertarian anarchy) does not necessarily foster or reward virtuous behavior, and that communitarian appeals (or, at their extreme, theocratic appeals) to entrenched values at the expense of freedom and spontaneity, republicanism offers an organizational vision that respects both flexibility and continuity. Republicanism explicitly endorses a sense of connectivity, a sense that any individual's freedom is undeniably dependent on her peer's freedoms. When freedoms are abused, exercised without responsibility, everyone's freedoms are threatened from within and without. Civic republicanism demands full engagement by all members of society. It values mutual respect and a recognition of common resources and common fates. Civic republicanism does not offer easy answers or a totalizing vision of the future. In fact, most republicans explicitly warn against such modes of thought and organization. In our human modesty, the best we can hope for is a sense that open, informed deliberation and general patience can yield better decisions and a common sense of the good life. In matters of technology, information, and communication, the need for a civic republican revival has never been so urgent.[5]

Those of us who are concerned about issues of open communication and information justice must generate richer, more informed debate on many fronts and across borders and oceans. We must invent global institutions that can facilitate deliberation and serve public needs. Republican activism is growing, but it is still fractured and inchoate. The anarchists might seem to be our allies. But they are better at uncivil disobedience than civic discussion. We would be better off with less disobedience and more deliberation.

ACKNOWLEDGMENTS

It takes a library to write a book. In this case, it took about a half dozen libraries. I used collections and services at New York University, the University of Wisconsin at Madison, the New York Public Library, the Buffalo and Erie County Public Libraries, the State University of New York at Buffalo, and that amazing tangled, anarchic, mess of a library that we call the Internet. So this book—and every book—owes its existence to the hundreds of thousands of professionals and volunteers around the world who strive against formidable forces to keep information flowing to those who need it most.

It also takes a community of generous friends and scholars to write a book. This book was inspired by one conference and solidified at another. In the spring of 2000 my dear friend Yochai Benkler brought scholars of law, communication, and economics together in New York City. Inspired by this conference, I started long, rich conversations with Michael Birnhack, James Boyle, Julie Cohen, Bernt Hugenholtz, Peter Jaszi, Rick Karr, Niva Elkin-Koren, Lawrence Lessig, Jessica Litman, Helen Nissenbaum, Gigi Sohn, and Alan Toner. The talks and papers from that conference inspired the vocabulary and agenda of questions and arguments that inspired this project. As I finished the manuscript in the spring of 2003, I had the honor and pleasure of spending a week at the Rockefeller Foundation center at Bellagio, Italy. Helen Nissenbaum and Monroe Price invited twenty-five experts, activists, and scholars from around the globe to discuss the ways technological standards and practices embody and influence values. Andrew Moss of Microsoft challenged me to defend my simplistic criticisms of certain convicted oligopolistic software corporations. Lawrence Liang inspired me to rethink my assumptions about the information ecosystems of the developing and urbanizing world. My colleague Alex Galloway—artist, coder, teacher, and scholar—joined me on that trip and served as an expert technological critic and supportive friend in the last months of this book's composition.

Between these landmark meetings, I contributed a little and learned much more from conferences and roundtable meetings sponsored by the Social Science Research Council, and the Rockefeller, Ford, and MacArthur Foundations. The Washington office of the American Library Association sponsored several talks and meetings at which I learned much. I delivered portions of this work at two meetings sponsored by the hacker magazine *2600*, a meeting of the Union for Democratic Communication, and two annual meetings of the American Library Association. Much of this book came from lectures I delivered at Stanford University Law School, the Wesleyan University music department, the Brown University music department, New York University School of Law, New York University School of Education, Cardozo Law School, the University at Buffalo School of Law, the offices of openDemocracy.net in London, the University of Wisconsin School of Library and Information Studies, the University of Illinois at Chicago Department of Communication, the University of Maryland University College, the University of Arizona Library School, the University of Georgia Humanities Center, the Communication program at the Massachusetts Institute of Technology, Harvard Law School, and Columbia University Law School.

I tried out some of the ideas in this book through articles I wrote for the *Chronicle of Higher Education*, *The Nation*, Salon.com, MSNBC.com, *New York Times Magazine*, *Academe*, and Blueear.com. The editors and publishers deserve applause for dealing with my tardy submissions and choppy prose. Eric Alterman has helped me reach bigger and smarter audiences by frequently linking his blog to mine. Above all, openDemocracy.net provided me with an ideal forum to audition these issues globally. Anthony Barnett, Paul Hilder, Solana Larson, and Casper Melville of openDemocracy.net edited deftly and allowed me to solicit instant feedback from a group of enlightened intellectuals around the world. Richard Barbrook, Miriam Rainsford, Matthew Rimmer, Roger Tatoud, Bill Thompson, and Sandy Starr all took the bait and delivered helpful and generous criticisms.

The ideas in this book should not seem new to participants in Slashdot.org and e-mail lists like cyberprof and Pho. I have lifted a lot from these communities. The chapter about music benefited greatly from conversations with Chuck D, John Flansburgh, Tim Quirk, Paul Miller, Dan O'Neill, Jenny Toomey, and Rob Walker. In fact, you could trace the germination of this book to a series of e-mails Rob Walker and I traded back in early 2000, as we were trying to make sense of that new thing called Napster.

My supportive colleagues at the University of Wisconsin, the University of Amsterdam, and New York University put up with all the distractions and deadlines that this book created for me. Ted Magder, Susan Murray, and Jay Rosen deserve special praise. Their patience and support is above and beyond the call of duty. Todd Gitlin moved to bigger digs uptown, but he never turned his back on his downtown friends. His counsel on many important matters has helped make my life and career richly rewarding.

My brilliant agent, Sam Stoloff, saw potential in me and this project back when it was a tangled assemblage of hunches and observations. Elizabeth Maguire at Basic Books was more patient with me than she probably should have been. William Frucht is a gentle editor and passionate polymath in a business in which such figures are increasingly rare.

I remain deeply indebted to Shelley Fisher Fishkin, the best mentor and role model a young scholar could hope for. My time in Austin, Texas, seems like such a long time ago, but friends from last century—especially Karen Adams, Joe Belk, Joel Dinerstein, Paul Erickson, Sue Krenek, Carolyn de la Peña, and Bob Randall—continue to inspire and drive me.

Some important voices went silent during the composition of this book: Johnny Cash, Jam Master Jay, Joey Ramone, Nina Simone, Joe Strummer, and Warren Zevon. This world is a poorer, less just, less fun place without them. The deepest, warmest voice that has gone silent recently belonged to Neil Postman. I will always treasure the moments I spent with him. He was the greatest teacher I have ever met. His humor and humane spirit live on through his family, students, and friends. We all miss him dearly.

As much as I appreciate my adopted family of friends and colleagues, my real family nurtures me more. My sisters cheered me on and my parents waited patiently for good news as I plodded through the writing of this book. My dear mother-in-law, Ann Henriksen, has made me feel loved in my new family of (gulp) Red Sox fans. My adorable stepdog, Eleanor Roosevelt Henriksen, sat quietly at my feet as I wrote.

Melissa Henriksen, my loving partner, deserves unlimited gratitude for taking a chance on a stray mutt like me. My favorite Warren Zevon song used to be "Accidentally Like a Martyr." Since Melissa came into my life, it's "Let Nothing Come Between You." To her, I dedicate this book and the rest of my life.

NOTES

Introduction

1. The concept of a moral panic is central to this argument. See Philip Jenkins, *Moral Panic: Changing Concepts of the Child Molester in Modern America* (New Haven, Conn.: Yale University Press, 1998); Jenkins, *Beyond Tolerance: Child Pornography on the Internet* (New York: New York University Press, 2001).

2. Howard Rheingold, *Smart Mobs: The Next Social Revolution* (Cambridge, Mass.: Perseus, 2002).

Chapter 1

1. Robert Darnton, "Paris: The Early Internet," *New York Review of Books,* June 29, 2000. Available at www.nybooks.com/articles/27.

2. Robert Darnton, *The Forbidden Best-Sellers of Pre-Revolutionary France* (New York: Norton, 1995). Also see Robert Darnton, *The Great Cat Massacre and Other Episodes in French Cultural History* (New York: Basic, 1984). Communicative anarchy mattered in the American Revolution as well. Darnton has been working in the same spirit of cultural history as American historian Bernard Bailyn, who won a Pulitzer Prize in 1968 for his book *The Ideological Origins of the American Revolution.* Bailyn took seriously the speeches and writings of American revolutionaries who were not admitted to the pantheon of American history. He wanted to find out what was on the mind of the English citizen-soldiers who in half a generation redefined themselves as Americans. See Bernard Bailyn, *The Ideological Origins of the American Revolution* (Cambridge, Mass.: Belknap Press, Harvard University Press, 1967).

3. Sean M. Sheehan, *Anarchism: Focus on Contemporary Issues* (London: Reaktion, 2003).

4. Robert Nozick, *Anarchy, State, and Utopia* (New York: Basic, 1974). Also see Robert Nozick and Jeffrey Paul, *Reading Nozick: Essays on Anarchy, State, and Utopia, Philosophy and Society* (Totowa, N.J.: Rowman & Littlefield, 1981).

5. G. E. Aylmer, *The Levellers in the English Revolution: Documents of Revolution* (Ithaca, N.Y.: Cornell University Press, 1975). Also see Todd Gitlin, *The Sixties: Years of Hope, Days of Rage* (Toronto: Bantam, 1987).

6. Sheehan, *Anarchism.*

7. Paul Avrich, *The Haymarket Tragedy* (Princeton, N.J.: Princeton University Press, 1984).

8. Eric Rauchway, *Murdering McKinley: The Making of Theodore Roosevelt's America* (New York: Hill & Wang, 2003).

9. Rauchway, *Murdering McKinley,* 77–79.

10. Matthew Frye Jacobson, *Barbarian Virtues: The United States Encounters Foreign Peoples at Home and Abroad, 1876–1917* (New York: Hill & Wang, 2000).

11. Peter Glassgold, *Anarchy! An Anthology of Emma Goldman's Mother Earth* (Washington, D.C.: Counterpoint, 2001); Emma Goldman and Alix Kates Shulman, *Red Emma Speaks: An Emma Goldman Reader* (New York: Schocken, 1983).

12. Sheehan, Anarchism. Also see Petr Alekseevich Kropotkin, *Anarchism and Anarchist Communism,* ed. Nicolas Walter (London: Freedom, 1987).

13. Sheehan, *Anarchism.* Also see George Orwell and Lionel Trilling, *Homage to Catalonia* (San Diego: Harcourt Brace, 1980).

14. Sheehan, *Anarchism.*

15. Geert Mak, *Amsterdam: A Brief Life of the City* (London: Harvill, 1999).

16. Don McNeill, *Moving Through Here* (New York: Knopf, 1970). Also see Gitlin, *The Sixties.*

17. Simson Garfinkel, "Leaderless Resistance Today," *First Monday* 8, no. 3 (2003).

18. Michel Foucault et al., *The Foucault Effect: Studies in Governmentality: With Two Lectures by and an Interview with Michel Foucault* (Chicago: University of Chicago Press, 1991). Also see Jim Miller, *The Passion of Michel Foucault* (New York: Simon & Schuster, 1993). Also see "Interview with Noam Chomsky on Anarchism, Marxism, and Hope for the Future," in Peter Ludlow, *Crypto Anarchy, Cyberstates, and Pirate Utopias,* Digital Communication (Cambridge, Mass.: MIT Press, 2001). Also see Chomsky, introduction to *Anarchism: From Theory to Practice* (New York: Monthly Review Press, 1970).

19. Sheehan, *Anarchism.*

20. Sheehan, *Anarchism.*

21. Robert D. Kaplan, *The Coming Anarchy: Shattering the Dreams of the Post Cold War* (New York: Random House, 2000). The book emerged from an article that Kaplan wrote for the *Atlantic Monthly* in 1995. See Robert D. Kaplan, "The Coming Anarchy," *Atlantic Monthly,* February 1994. The strongest response to Kaplan, who did little more than offer a catalog of horrors and excesses as snapshots that somehow add up to a trend or movement, is Jennifer Widner, "States and Statelessness in Late Twentieth-Century Africa," *Daedalus,* Summer 1995.

22. Nozick, *Anarchy, State, and Utopia.* Also see Nozick and Paul, *Reading Nozick.*

23. See Kevin Doyle, "Interview with Noam Chomsky on Anarchism, Marxism, and the Hope for the Future," in Ludlow, *Crypto Anarchy, Cyberstates, and Pirate Utopias.*

Chapter 2

1. Rachel Lehmann-Haupt, "The Music Pirates," November 1997. Available at www.wired.com/wired/archive/5.11/eword.html?pg=6.

2. Clay Shirky, "What Is P2P . . . and What Isn't?" *O'Reilly Network*, November 24, 2000. Available at www.openp2p.com/pub/a/p2p/2000/11/24/shirky1-whatisp2p .html

3. Distributed systems are not necessarily emergent systems, but some of them could be. As Stephen Johnson explains in his brilliant book *Emergence*, emergent systems are self-taught. They evolve. They have feedback protocols built into their design. So they learn from experience. The systems I am considering might loosely resemble emergent systems, and some might even be emergent, but I would caution against simple general conflation. Emergence can look like a process of simulated centralization or progressive uniformity. Johnson clearly explains that the Internet is not an emergent system but elements of it behave along emergent lines. The editing and Karma protocols on dynamic Web sites like Slashdot.org are the best examples of this phenomenon. But the Internet as it currently operates, based on TCP/IP, does not learn. Despite physicist Freeman Dyson's predictions, the Internet is not a giant global brain. And that's good—because it would be a pretty filthy brain. See Stephen Johnson, *Emergence* (New York: Scribners, 2001).

4. Paul T. Durban, "Philosophy of Technology: Retrospective and Prospective Views," in Eric Higgs, Andrew Light, and David Strong, eds., *Technology and the Good Life?* (Chicago: University of Chicago Press, 2000).

5. Marshall McLuhan, *Understanding Media: The Extensions of Man* (Cambridge, Mass.: MIT Press, 1994).

6. Jack M. Balkin, *Cultural Software: A Theory of Ideology* (New Haven, Conn.: Yale University Press, 1998).

7. See McLuhan, *Understanding Media*. Also see Neil Postman, *Technopoly: The Surrender of Culture to Technology* (New York: Vintage, 1993). For the quote from the Diggers, see Gitlin, *The Sixties*.

8. Thorstein Veblen, *The Theory of the Leisure Class; an Economic Study of Institutions*, introduction by C. Wright Mills (New York: New American Library, 1959).

9. Pierre Lévy, *Collective Intelligence: Mankind's Emerging World in Cyberspace* (Cambridge, Mass.: Perseus, 1999). Also see Robert Wright, *Nonzero: The Logic of Human Destiny* (New York: Vintage, 2001).

10. Jessica Litman, *Digital Copyright: Protecting Intellectual Property on the Internet* (Amherst, N.Y.: Prometheus, 2001), p. 151. For an argument in favor of global market fundamentalism, see George F. Gilder, *The Spirit of Enterprise* (New York:

Simon & Schuster, 1984). Also see Thomas L. Friedman, *The Lexus and the Olive Tree* (New York: Anchor, 2000). For a critique of market fundamentalism, see Thomas Frank, *One Market Under God: Extreme Capitalism, Market Populism, and the End of Economic Democracy* (New York: Doubleday, 2000).

11. Siva Vaidhyanathan, *Copyrights and Copywrongs: The Rise of Intellectual Property and How It Threatens Creativity* (New York: New York University Press, 2001).

12. An example of this concerns the use of globalizing information technology by antiglobalization activists. See Jackie Smith, "Cyber Subversion in the Information Economy," *Dissent*, Spring 2001, pp. 48–52.

Chapter 3

1. Robert Bracht Branham and Marie-Odile Goulet-Cazé, *The Cynics: The Cynic Movement in Antiquity and Its Legacy*, Hellenistic Culture and Society, No. 23 (Berkeley: University of California Press, 1996). This collection of essays by scholars of the ancient world is the most important source of my understanding of the Cynics and their teachings. Martha Nussbaum's essays on cosmopolitanism stimulated my interest in cynicism and its potential value in my teaching and research. See Martha Nussbaum, "Patriotism and Cosmopolitanism," in Martha Craven Nussbaum and Joshua Cohen, *For Love of Country: Debating the Limits of Patriotism* (Boston: Beacon, 1996). Also see Nussbaum, *Cultivating Humanity: A Classical Defense of Reform in Liberal Education* (Cambridge, Mass.: Harvard University Press, 1997); Orlando Patterson, *Freedom in the Making of Western Culture* (New York: Basic, 1991).

2. Diogenes is a registered trademark of the joint venture DIOGENES, composed of FOI Services, Inc., and Washington Business Information, Inc. It is the name of a database service that contains full text and citations of the U.S. Food and Drug Administration regulatory information needed by the health care industry. See STN Database Summary Sheet, www.cas.org/DBSS/diogenesss.html.

3. See Pew Internet and American Life Project, Daily Internet Activities (December 2001) at www.pewinternet.org/reports.

4. Thomas Friedman, "Webbed, Wired, and Worried," *New York Times*, May 26, 2002, p. 11.

5. For a revealing profile of Clarke, see Declan McCullagh, "The Sentinel," *Wired*, March 2002, 105–107.

6. www.info-sec.com/ciao/bioclarke.html

7. Richard Clarke, "New Ideas to Advance the Network Economy" (speech presented to the global tech summit, sponsored by the Business Software Alliance in Washington, D.C., December 4, 2001).

8. Gary C. Kessler, "An Overview of TCP/IP Protocols and the Internet," http://compnetworking.about.com/gi/dynamic/offsite.htm?site=http%3A%2F%2F www.garykessler.net%2Flibrary%2Ftcpip.html.

9. My understanding of the distinction between protocols and controls is not shared by some scholars of culture and technology whom I deeply respect. My colleague Alex Galloway, for instance, offers a much more complex description of how protocols work. He differs with me on diction and how to judge the impacts of protocols. Galloway asserts that protocols are a type of control, explaining, as Michel Foucault or Gilles Deleuze might say, that prisons are "confining" instead of "controlling." For a brilliant account of protocols and their effects, see Alexander R. Galloway, *Protocol: How Control Exists After Decentralization* (Cambridge, Mass.: MIT Press, 2004). Also see Gilles Deleuze, *Negotiations, 1972-1990: European Perspectives* (New York: Columbia University Press, 1995). Also see Michel Foucault, *Discipline and Punish: The Birth of the Prison,* 2d ed. (New York: Vintage, 1995).

10. Tim Berners-Lee and Mark Fischetti, *Weaving the Web: The Original Design and Ultimate Destiny of the World Wide Web by Its Inventor* (New York: Harper-Collins, 2000).

11. Lawrence Lessig, *The Future of Ideas: The Fate of the Commons in a Connected World* (New York: Random House, 2001).

12. A plaintiff in a copyright suit need not prove actual damages caused by infringement; the act of infringement presumes harm. To justify an injunction, a plaintiff must convince a court that it would suffer irreparable harm if the allegedly infringing action were to continue.

13. For an explanation of why Spar's historicist and inductive intellectual framework easily collapses under scrutiny, see Karl Raimund Popper, *The Poverty of Historicism* (Boston: Beacon, 1957). Also see Karl Raimund Popper, *The Logic of Scientific Discovery* (London: Routledge, 1992).

14. Debora L. Spar, *Ruling the Waves: Cycles of Discovery, Chaos, and Wealth from the Compass to the Internet* (New York: Harcourt, 2001).

15. For the best account of the evolution of ICANN and the domain name system, see Milton Mueller, *Ruling the Root: Internet Governance and the Taming of Cyberspace* (Cambridge, Mass.: MIT Press, 2002).

16. See Jonathan Zittrain, "Book Review: Ruling the Root," *Federal Communications Law Journal,* December 2002.

Chapter 4

1. For Kazaa software installation figures, see www.kazaa.com/us/news/most_downloaded.htm. Compact disc sales and revenue figures come from the Recording Industry Association of America (RIAA) and are available at www.riaa.com. The figures for 2002 come from a variety of news accounts, including Damien Cave, "File Sharing: Guilty as Charged," in Salon.com, August 23, 2002. Also see Laura M. Holson, "Twilight of the CD? Not if It Can Be Reinvented," *New York Times,* February 23, 2003. Many news reports measure the financial changes in the music industry in

confusing ways. There seems to be no industry standard for measuring sales or rates of change in sales. Compact disc sales figures are not ideal measures of the health of the industry. These RIAA numbers come from a service called Soundscan, which does not include sales from small record stores, Internet sales (Amazon.com, etc.), authorized Internet downloading sites (Apple iTunes, Pressplay, etc.), or mail-order CD clubs (Columbia House, BMG Music Club). The industry is certainly manipulating the numbers to affect the terms of public debate on the issues.

2. The Napster trademark is now owned by Roxio, a company that specializes in CD burning and multimedia software. In the fall of 2003 Roxio launched Napster 2.0, a controlled digital download system with licensed content streamed from a central server. Nothing about the new Napster resembles the old.

3. Frank Rose, "The Civil Way Inside Sony," *Wired,* February 2003.

4. "A Big Fat Thanks to Record Execs," *New York Times,* October 28, 2002. C6. For an account of the development and dynamics of subcultures, see Dick Hebdige, *Subculture: The Meaning of Style,* New Accents (London: Routledge, 1991); Hebdige, *Cut 'N' Mix: Culture, Identity, and Caribbean Music* (London: Methuen, 1987); Tricia Rose, *Black Noise: Rap Music and Black Culture in Contemporary America,* Music/Culture (Hanover, N.H.: Wesleyan University Press/University Press of New England, 1994); Andrew Ross and Tricia Rose, *Microphone Fiends: Youth Music and Youth Culture* (New York: Routledge, 1994). For a succinct account of the development of hip-hop, see Siva Vaidhyanathan, *Copyrights and Copywrongs: The Rise of Intellectual Property and How It Threatens Creativity* (New York: New York University Press, 2001).

5. "Most Wanted," *New York Times,* October 28, 2002. C6. The ranks come from SoundScan, the music industry's point-of-sale market research company.

6. Josh Grossberg, "Eminem Show Opening Earlier," E Online, May 24, 2002. Available at www.eonline.com/News/Items/0,1,10002,00.html.

7. John Borland, "Eminem CD Spotlights New Piracy Patterns," News.com, May 28, 2002. Available at news.com.com/2100-1023-923472.html.

8. "Eminem's 'The Eminem Show' Album Tops Weekly Sales Chart," Reuters News Service, June 12, 2002. Available at http://ca.news.yahoo.com/020612/5/n0yj.html. Also see Justin Oppelaar, "Eminem 'Show' Top Seller of 2002," MSNBC.com, December 27, 2002. Available at www.msnbc.com/news/852342.asp.

9. For an account of how the commercial music industry and the commercial radio industry in the United States is exploiting its labor force, see the Future of Music Coalition, www.futureofmusic.org.

10. See www.forrester.com/ER/Research/Report/Summary/0,1338,14854,FF.html. Also see Andrew Orlowski, "Missing RIAA figures shoot down 'piracy' canard," *Register,* December 16, 2002. Available at www.theregister.co.uk/content/6/28588.html.

11. Robert Wilonsky, "Do the Math: The Music Industry Says Online Piracy's Killing the Biz. A UTD Prof Says It Ain't," *Dallas Observer,* August 8, 2002. Available at www.dallasobserver.com/issues/2002-08-08/stuff.html/1/index.html. Also see

Damien Cave, "File Sharing: Innocent Until Proven Guilty," Salon.com, June 13, 2002; Cave, "File Sharing: Guilty As Charged," Salon.com, August 23, 2002. Generally, see S. J. Liebowitz, *Re-Thinking the Network Economy: The True Forces That Drive the Digital Marketplace* (New York: Amacom, 2002). Also see the addendum and update to Stan Liebowitz, *Record Sales, MP3 Downloads, and the Annihilation Hypothesis*. Available at www.utdallas.edu/~liebowit/knowledge_goods/records.pdf.

12. For the best argument for and description of the jukebox, see Paul Goldstein, *Copyright's Highway: From Gutenberg to the Celestial Jukebox* (New York: Hill & Wang, 1994). Regarding the case for efficient digital content delivery, see Nicholas Negroponte, *Being Digital* (New York: Knopf, 1995). For a critique of Negroponte and the antidemocratic potential of the targeted, filtered, efficient delivery of information, see Cass R. Sunstein, *Republic.Com* (Princeton, N.J.: Princeton University Press, 2001).

13. http://www.sdmi.org

14. John Alderman, *Sonic Boom: Napster, Mp3, and the New Pioneers of Music* (Cambridge, Mass.: Perseus, 2001), p. 91. Other important books on the peer-to-peer phenomena in music include Michael Lewis, *Next: The Future Just Happened* (New York: Norton, 2001). The best brief account of the effects of peer-to-peer systems on copyright law and vice versa is Kathy Bowrey and Matthew Rimmer, "Rip, Mix, Burn: The Politics of Peer to Peer and Copyright Law," *First Monday* 7, no. 8 (2002).

15. Alderman, *Sonic Boom*, p. 91.

16. Michael Learmonth, "AOL and InterTrust: A Legal Napster," *Industry Standard*, July 10, 2000. Available at www.thestandard.com/article/display/0,1151,16564,00.html.

17. See Vaidhyanathan, *Copyrights and Copywrongs*.

18. Alderman, *Sonic Boom*, p. 27. Engineers working for the German research company Fraunhofer developed the MP3 protocols. Fraunhofer patented and licensed the algorithm.

19. For information on the battle between Felten and the RIAA, see www.cs.princeton.edu/sip/sdmi/faq.html#A0.

20. Letter from Matthew J. Oppenheim to Professor Edward Felten, April 9, 2001. Available at www.cs.princeton.edu/sip/sdmi/riaaletter.html.

21. "Toto Recall," *Harper's Magazine*, March 2003, 20.

22. Nostalgic for the working zipper on *Sticky Fingers*, I fought compact discs as long as I could. The re-release of the Beach Boys' *Pet Sounds* in 1988 won me over. It was and is brilliant and life changing. I bought my first CD player soon after *Pet Sounds* came out again. It remains the best album ever made. I was pleased to find that others remember vinyl as fondly as I. Matthew Sweet included the end-of-side "shooshes" halfway through his compact disc of his album *Girlfriend*.

23. Holson, "Twilight of the CD?"

24. Scott Carlson, "Recording Industry Plans to Accelerate Complaints About Illegal File Sharing," *Chronicle of Higher Education*, January 3, 2003, A38.

25. Dan Ackman, "Record Industry Is Taking Names," Forbes.com, January 27, 2003.

26. Jane Black, "The Keys to Ending Music Piracy," Business Week Online, January 27, 2003.

27. Kelly Donahue, "Musicnet and Pressplay: To Trust or Antitrust," *Duke Law and Technology Review,* November 12, 2001. Available at www.law.duke.edu/journals /dltr/articles/2001dltr0039.html. As of April 2003, neither PressPlay nor MusicNet was usable by computers that use Linux or Macintosh operating systems.

28. Declan McCullagh, "Perspective: The New Jailbird Jingle," News.com, January 27, 2003.

29. Siva Vaidhyanathan, "MP3: It's Only Rock and Roll and the Kids are Alright," *Nation,* July 24, 2000.

30. Randy Cohen, *The Good, the Bad and the Difference: How to Tell Right from Wrong in Everyday Situations* (New York: Doubleday, 2002). Sometime after our debate occurred privately, via e-mail, Cohen asked for permission to reprint my comments in his book. I replied with an "emoticon" chuckle that he need not have asked, since I consider the use of my copyrighted e-mails to be fair use. I have reciprocated by quoting from his book in mine. This is how conversation happens. Excessive enforcement of copyright or denial of fair use would have raised the price of this conversation above the level I would pay.

Chapter 5

1. Amy Harmon, "Music Industry in Global Fight on Web Copies," *New York Times,* October 7, 2002.

2. See MPAA, "Valenti Announces Dramatic Box Office and Admissions Increases in Decades in Showest Address," March 4, 2003. Available at www.mpaa.org/ jack/2003/2003_03_05A.htm.

3. The American motion picture industry estimates 2002 piracy losses as $3 billion. It's entirely unclear how the MPAA generates this number. See www.mpaa.org/ anti-piracy.

4. The MPAA estimates it loses more than $3 billion per year to simple analog videotape piracy. See www.mpaa.org/jack/2002/2002_02_12b.htm.

5. In 2002 the Motion Picture Association of America reported its members enjoyed box office totals of $9.5 billion, an increase of 13.2 percent over 2001 and the highest year-to-year increase in twenty years. The MPAA also announced that in 2002 admissions overseas totaled 7.3 billion, an increase of 7.5 percent from the previous year. See www.mpaa.org/jack/2003/2003_03_05A.htm.

6. See John Healey, "Success of DVD Players Proves a Mixed Blessing," *Los Angeles Times,* January 6, 2003. Available at www.latimes.com/la-fi-dvd6jan06001441, 0,2780651.story. Also see "DVD Sales Double in 2001," BBC News, January 4, 2002. Available at http://news.bbc.co.uk/1/hi/entertainment/new_media/1742959.stm.

7. Lawrence Lessig, *The Future of Ideas: The Fate of the Commons in a Connected World* (New York: Random House, 2001).

8. See Patricia Aufderheide, *Communications Policy and the Public Interest*, Guilford Communication Series (New York: Guilford, 1999).

9. Mike Godwin, "Prisoners of Digital Television," Reason Online, April 2003.

10. See Marjorie Heins, *The Progress of Science and Useful Arts: Why Copyright Today Threatens Intellectual Freedom* (New York: Free Expression Policy Project, 2003). For the text of the Hollings bill, see www.politechbot.com/docs/cbdtpa/hollings.s2048.032102.html. For background, see www.politechbot.com/docs/cbdtpa. For the list of absurd applications of this proposal, see www.freedom-to-tinker.com.

11. Daniel Kraus, "The Phantom Edit," Salon.com, November 5, 2001. Available at http://archive.salon.com/ent/movies/feature/2001/11/05/phantom_edit/. Also see Peter Rojas, "The Blessed Vision," *Village Voice*, October 9, 2002. Available at www.villagevoice.com/issues/0241/rojas.php.

12. Judith Crosson, "Hollywood Studios Join Film-Sanitizing Lawsuit," Forbes.com, December 13, 2002. Also see Drew Clark, "Bowdlerizing for Columbine: Why American Directors Have No Moral Rights to Their Movies," *Slate*, January 20, 2003. Available at http://slate.msn.com/id/2077192.

13. Nick Paton Walsh, "Russia's Cult Video Pirate Rescripts Lord of the Rings As Gangster Film," *Observer*, June 22, 2003.

Chapter 6

1. Alice Randall, *The Wind Done Gone* (Boston: Houghton Mifflin, 2001).

2. See Lawrence Lessig, "Let the Stories Go," *New York Times*, April 30, 2001, p. A19.

3. Siva Vaidhyanathan, *Copyrights and Copywrongs: The Rise of Intellectual Property and How It Threatens Creativity* (New York: New York University Press, 2001), pp. 145–148.

4. *MCA, Inc. v. Wilson*, 677 F, 2d 180 (2d Cir. 1981).

5. See Jed Rubenfeld, "The Freedom of Imagination: Copyright's Constitutionality," *Yale Law Journal*, volume 112.

6. Vaidhyanathan, *Copyrights and Copywrongs*.

7. See Steve Seidenberg, "Copyright Owners Sue Competitors," *National Law Journal*, February 3, 2003. Available at www.nlj.com/business/020303bizlede.shtml.

8. The best account of recent American battles over intellectual property, with copyright at its center, is Lawrence Lessig, *The Future of Ideas: The Fate of the Commons in a Connected World* (New York: Random House, 2001). A somewhat less accessible yet more profound account of the ways laws and computer code are steadily regulating information and creativity can be found in Lawrence Lessig, *Code and Other Laws of Cyberspace* (New York: Basic, 1999). For the ways that changes in technology are complicating traditional legal systems, see Stuart Biegel, *Beyond Our Control? Confronting the Limits of Our Legal System in the Age of Cyberspace*

(Cambridge, Mass.: MIT Press, 2001). An excellent account of recent changes in copyright since the rise of digital technology is Jessica Litman, *Digital Copyright: Protecting Intellectual Property on the Internet* (Amherst, N.Y.: Prometheus, 2001). Two recent books consider the long sweep of copyright law. Celebrating efforts to expand protection is Edward B. Samuels, *The Illustrated Story of Copyright* (New York: Thomas Dunne/St. Martin's, 2000). More critical of recent trends is Vaidhyanathan, *Copyrights and Copywrongs.* For an account of the development of patent law, few sources are written for the nonspecialist. Two essential sources, both highly critical of recent efforts to expand patent protection, are Michael Perelman, *Steal This Idea: Intellectual Property Rights and the Corporate Confiscation of Creativity* (New York: Palgrave, 2002) and Seth Shulman, *Owning the Future* (Boston: Houghton Mifflin, 1999). Also see David Bollier, *Silent Theft: The Private Plunder of Our Common Wealth* (New York: Routledge, 2002). Keith Maskus presents a readable account of debates over the globalization of intellectual property regimes in *Intellectual Property Rights in the Global Economy* (Washington, D.C.: Institute for International Economics, 2000). The best books on the globalization of intellectual property regimes are Susan K. Sell, *Power and Ideas: North-South Politics of Intellectual Property and Antitrust,* Suny Series in Global Politics (Albany: State University of New York Press, 1998); Sell, *Private Power, Public Law: The Globalization of Intellectual Property Rights,* Cambridge Studies in International Relations, no. 88 (Cambridge: Cambridge University Press, 2003). Also see Christopher May, *A Global Political Economy of Intellectual Property Rights: The New Enclosures?* Routledge/Ripe Studies in Global Political Economy (London: Routledge, 2000). One cannot grasp the foundations of intellectual property without Peter Drahos, *A Philosophy of Intellectual Property* (Aldershot, U.K.: Dartmouth, 1996).

9. Vaidhyanathan, *Copyrights and Copywrongs.*

10. May, *Global Political Economy?* Sell, *Private Power.*

11. See Reginald Walter Michael Dias and Graham Hughes, *Jurisprudence* (London: Butterworth, 1957); Orlando Patterson, *Slavery and Social Death: A Comparative Study* (Cambridge, Mass.: Harvard University Press, 1982).

12. Jack Valenti, "A Clear and Future Danger," openDemocracy.net, May 30, 2002. Available at www.opendemocracy.net/debates/article.jsp?id=8&debateId=40&articleId=58.

13. Richard Stallman, "Let's Share!" openDemocracy.net, May 30, 2002. Available at www.opendemocracy.net/debates/article.jsp?id=8&debateId=40&articleId=31.

14. See L. Ray Patterson, *Copyright in Historical Perspective* (Nashville, Tenn.: Vanderbilt University Press, 1968); L. Ray Patterson and Stanley W. Lindberg, *The Nature of Copyright: A Law of Users' Rights* (Athens: University of Georgia Press, 1991).

15. See Vaidhyanathan, *Copyrights and Copywrongs.* In the United States, we have enjoyed a profound limitation on the power of the state to regulate speech. This limitation has its roots in the First Amendment to the U.S. Constitution, which forbids Congress from making laws impeding freedom of speech, press, or religion. This

freedom has evolved over time. At first it was almost never invoked, and rarely taken seriously. It has never been absolute. Only recently has it been incorporated into most areas of public discourse (state governments could censor well into the twentieth century). Now, more than two hundred years after the ratification of the First Amendment, the state has a very high burden to surmount if it wants to restrict the speech of any American. Copyright is the glaring exception. Courts have justified evading the tough question of the censorious power of copyright by maintaining blind faith in four limitations on the power of copyright. These four limitations are the source of what we might call "users' rights," which abut and complicate the copyright holder's rights.

The principle that after the "first sale" of a copyrighted item, the buyer can do whatever she wants with the item—save publicly performing the work or distributing unauthorized copies of it for sale. The first sale doctrine is what makes the lending library possible. The idea behind this concept is that ownership of the item is distinct from control of the copyright over the text, signals, or digits.

- The concept that copyright protects specific expression of ideas but not the ideas themselves. This is the least understood but perhaps most important tenet of copyright. You can't copyright a fact or an idea. Because you can't, anyone may repeat your idea to criticize it or build on it. Journalism, along with many other forms of common expression, relies on this principle.

- The promise that copyright would only last, as the Constitution demands, "for limited times," thus constantly replenishing the public domain. The public domain allows for low-cost scholarship, research, and revision of formerly copyrighted works. The reason that bookstores are filled with high-quality yet affordable scholarly editions of Mark Twain's *Adventures of Huckleberry Finn* and John Stuart Mill's *On Liberty* is that they are in the public domain. The reason there is no annotated scholarly edition of Ralph Ellison's *Invisible Man* is that it is not. The public domain is the essential "balloon payment" in the copyright deal. The public gives up the right to make unlimited and unregulated copies of material in exchange for the very existence of industries that will fill our shelves with books and our lives with Muzak. Eventually the public is paid back for enduring years of monopoly prices. It gets to use and reuse that work freely—for no cost and with no restrictions.

- The principle of fair use—at its base a legal defense against an accusation of copyright infringement. If you are accused of infringing, you can make an argument that your use of the protected works is "fair" because of some combination of the following four factors: the nature of the original work is important to public discussions or concerns; the nature of your use of it is important because of teaching, research, or commentary; you did not use very much of the original work; your use did not significantly affect the market for the original work. In the public discourse about fair use, it has served as a term representing a collection of uses that consumers could consider "fair," such as recording television shows for later viewing, making cassette tape or MP3 mixes from compact discs, and limited copying for private, noncommercial sharing.

Despite being the most widely used example of "content-based regulation," copyright cases do not invoke First Amendment scrutiny. Courts regularly issue injunctions (prior restraint) in copyright cases, before any evidentiary findings. See Rubenfeld, "Freedom of Imagination." Also see Mark Lemley and Eugene Volokh, "Freedom of Speech and Injunctions in Intellectual Property Cases," *Duke Law Journal*, November 1998.

Chapter 7

1. "Delhi Police Conducts Largest Piracy Raid," *India Infoline*, May 31, 2002. Available at www.Indiainfoline.com/news/news.asp?dat=3158.

2. Sudashana Banerjee, "Pirated Software up for Grabs," *Ciol*, July 29, 2002. Available at www.ciol.com/content/news/trends/102072901.asp.

3. My understanding of the piracy markets in India comes from several sources. First, I have some personal experience with pirated book and cassette tape bazaars in Madras from the late 1970s through the mid-1990s, before the Internet and easy optical disc replication changed everything. Second, I have been following the issues in several Indian publications. See "The Media Fabric of the Contemporary City: Publics and Practices in the History of the Present," *Broadsheet*, Spring 2003. Also see Sudashana Banerjee, "Pirated Software up for Grabs," *Ciol*, July 29, 2002. Available at www.ciol.com/content/news/trends/102072901.asp. Also see "Pirates of New Delhi," *Publishing Trends*, November 2001. The most important textual source about the changes in law and culture in India since the rise of the digital economy is Lawrence Liang, "Global Commons, Public Space and Contemporary IPR," World Association for Christian Communication. Available at www.wacc.org.uk/publications /md/md2003-1/lawrence.html. Also see Lawrence Liang and Sudhir Krishnaswamy, "IPR and the Knowledge/Culture Commons," www.sarai.net. I have learned most of what I know about copyright in India from conversations with Lawrence Liang.

4. Liang, "Global Commons, Public Space, and Contemporary IPR."

5. "Clouds Hang over Bollywood," BBC News, October 21, 2002. Available at http://news.bbc.co.uk/1/hi/business/2347339.stm. Here are some interesting facts about the Indian film industry: Since 1931 India has produced more than 67,000 films in more than thirty different languages and dialects. In 2001 the industry produced 1,013 films, making it the world's largest feature film producer. The majority of films are made in the South Indian languages of Telugu, Tamil, and Malayalam (537, compared to 230 in Hindi), but Hindi-language films take the largest box office share. The industry draws its revenues from domestic theatrical sales (2001: 36 billion rupees); overseas rights (2001: 5.25 billion rupees); music rights (2001: 1.5 billion rupees); television and video rights (2001: 2 billion rupees); corporate spon-

sorship and merchandising (2001: 10 million rupees).Total industry revenues from these sources are estimated at 45 billion rupees. India has the highest entertainment tax in the world, so the 36 billion rupees generated in 2001 through domestic theatrical sales equates to 72 billion rupees before taxes. The overseas market produces 25–30 percent of the total proceeds. Indian film exports increased by 17 percent, from 4.5 billion rupees in 2000 to 5.25 billion rupees in 2001. The United States, Canada, and the United Kingdom are the major export destinations. Other emerging markets include Japan, South Africa, Mauritius, Australia, New Zealand, and the Middle East. See www.filmcouncil.org.uk/filmindustry/?p=India.

6. Brian Larkin, "Degrading Images, Distorted Sounds: Nigerian Video and the Infrastructure of Piracy," *Public Culture* 16, no. 3 (2003). Also see Brian Larkin, "Bollywood Comes to Nigeria," Samar 8 (1997).

7. Graham Gori, "In Mexico, Pirated Music Outsells the Legal Kind," *New York Times*, April 1, 2002.

8. Eileen Southern, *The Music of Black Americans: A History*, 2d ed. (New York: Norton, 1983). Also see Christopher Small, *Music of the Common Tongue: Survival and Celebration in African American Music*, Music/Culture (Hanover, N.H.: University Press of New England, 1998). For an account of the social pressures that severed and splintered slave society, see Orlando Patterson, *Slavery and Social Death: A Comparative Study* (Cambridge, Mass.: Harvard University Press, 1982).

9. Stephen Feld, "A Sweet Lullaby for World Music," in Arjun Appadurai, *Globalization, A Millenial Quartet Book* (Durham, N.C.: Duke University Press, 2001).

10. Feld, "Sweet Lullaby," p. 167.

11. Feld, "Sweet Lullaby," pp. 167–168.

12. Matthew Arnold and Stefan Collini, *Culture and Anarchy and Other Writings: Cambridge Texts in the History of Political Thought* (Cambridge: Cambridge University Press, 1993). Also see John Stuart Mill, *On Liberty*, ed. David Spitz, Norton Critical Edition (New York: Norton, 1975). Also see Lionel Trilling, *Matthew Arnold* (New York: Norton, 1939).

13. Samuel Huntington, *The Clash of Civilizations and the Remaking of World Order* (New York: Touchstone, 1997), p. 58.

14. Tyler Cowen, *Creative Destruction: How Globalization Is Changing the World's Cultures* (Princeton, N.J.: Princeton University Press, 2002).

15. Lawrence W. Levine, *Black Culture and Black Consciousness: Afro-American Folk Thought from Slavery to Freedom* (New York: Oxford University Press, 1977).

16. Ted Magder, Canada's Hollywood: *The Canadian State and Feature Films*, The State and Economic Life, no. 18 (Toronto: University of Toronto Press, 1993).

17. Ted Magder, *Franchising the Candy Store: Split-Run Magazines and a New International Regime for Trade in Culture*, Canadian-American Public Policy, no. 34 (Orono, Me.: Canadian-American Center University of Maine, 1998).

18. The emerging study of cultural policy is contributing to our understanding of how cultures work in a dynamic economy and a minimal state. See Gigi Bradford et

al., *The Politics of Culture: Policy Perspectives for Individuals, Institutions, and Communities* (New York: New Press/Norton, 2000). Many scholars today are interested in issues of cultural policy, for example, Russell Cargo, founder and director of the nonprofit management program at George Mason University; Margaret Wyszomirski, director of the arts policy and administration program at Ohio State University; and George Yudice, director of the privatization of culture project at New York University. Communications scholars also do interesting work in this area (although they rarely employ the phrase "cultural policy" to describe what they do). Robert McChesney, University of Illinois at Urbana-Champaign, has written about the ways certain broadcasting companies rigged the federal regulating game in the 1930s and retarded the growth of other companies. At NYU, Mark Crispin Miller has run the Project on Media Ownership (PROMO), which analyzes the ways that media mergers have corrupted the public sphere. The list could go on. To date, however, this work hasn't often been integrated into schools of policy studies. Neither the John F. Kennedy School of Government at Harvard University nor the Lyndon B. Johnson School of Public Affairs at the University of Texas at Austin has anyone on the permanent faculty, as far as I know, devoted to anything like "cultural policy." At the University of Chicago, in contrast, the Harris School of Public Policy Studies has teamed up with the humanities division to establish a cultural policy program directed by Lawrence Rothfield, an associate professor of English and comparative literature. Another notable center of study is at Princeton University, where Stanley N. Katz, former president of the American Council of Learned Societies, is a professor at the Woodrow Wilson School of Public and International Affairs and director of the Princeton University Center for Arts and Cultural Policy Studies. Some of the most detailed work on cultural policy is being conducted in the world of private philanthropy. In 1998 a consortium of foundations set up the Center for Arts and Culture in Washington, which bills itself as "the first independent think tank for arts and cultural issues." The center has established a cultural policy network to link hundreds of scholars in disparate academic enclaves. Support from such institutions is cause for optimism: although cultural policy remains justifiably ad hoc, the study of it is not.

Chapter 8

1. Nehad Selaiha, "A Play for Today," *Al-Ahram Weekly* On-line, December 22–January 2, 2002. Selaiha wrote, "A triumphant Galileo would not make any sense to an Egyptian, would seem utterly divorced from reality—a facile fabrication of wishful thinking. His tragic descent from bright optimism to disillusion, from proud defiance to servility and humiliation and from honourable courage to cowardly betrayal is an experience too familiar to many Egyptians not to touch a raw nerve somewhere." Also see Selaiha, "Brecht in Egypt," *Al-Ahram Weekly* On-line, June 11–17, 1998.

2. Ravi Vasudevan, Geert Lovink, Lawrence Liang, Ravi Sundaram, and others have been exploring the social dynamics of "illegal cities," urban networks of people who dodge in and out of legal regimes in the overflowing cities of the developing world. See Centre for the Study of Developing Societies and Society for Old and New Media (Amsterdam, Netherlands), *The Cities of Everyday Life*, Sarai Reader, no. 2 (New Delhi: Centre for the Study of Developing Societies/Society for Old and New Media/Rainbow Publishers, 2002).

3. Elise Ackerman and Joshua L. Kwan, "Egypt Aims to Enter Internet Age in One Big Step," *San Jose Mercury-News,* December 1, 2002. Available at www.siliconvalley.com/mld/siliconvalley/4641662.htm?template=contentModules/printstory.jsp.

4. Sue Anne Pressley and Justin Blum, "Hijackers May Have Accessed Computers at Public Libraries; Authorities Investigating Possible Internet Communications," *Washington Post,* September 16, 2001.

5. *Uniting and Strengthening America by Providing Appropriate Tools Required to Intercept and Obstruct Terrorism (USA Patriot Act) Act of 2001*, 107th Congress, 1st sess., H.R. 3162. Also see Nancy Chang, *Silencing Political Dissent* (New York: Seven Stories, 2002); Alphonse B. Ewing, *The USA Patriot Act* (New York: Novinka, 2002); Genevieve Johanna Knezo, *Federal Security Controls on Scientific and Technical Information: History and Current Controversy* (New York: Nova Science, 2003); Bernard D. Reams and Christopher Anglim, *USA Patriot Act: A Legislative History of the Uniting and Strengthening of America by Providing Appropriate Tools Required to Intercept and Obstruct Terrorism Act, Public Law no. 107-56 (2001)*, 5 vols. (Buffalo, N.Y.: Hein, 2002); Regulatory Compliance Associates, Inc., and Sheshunoff Information Services, *Special Report, USA Patriot Act: A Guide to Regulatory Compliance* (Austin, Tex.: Thomson/Sheshunoff, 2002); House Committee on Financial Services, *Terrorist Financing: Implementation of the USA Patriot Act: Hearing Before the Committee on Financial Services, U.S. House of Representatives*, 107th Cong., 2d sess., September 19, 2002 (Washington, D.C.: GPO, 2002); Senate Committee on Banking, Housing, and Urban Affairs, *The Financial War on Terrorism and the Administration's Implementation of Title iii of the USA Patriot Act: Hearing Before the Committee on Banking, Housing, and Urban Affairs, United States Senate, 107th Cong., 2d sess., on the Administration's Implementation of the Anti-Money Laundering Provisions (Title iii) of the USA Patriot Act (Public Law 107-56), and Its Efforts to Disrupt Terrorist Financing Activities, January 29, 2002* (Washington, D.C.: GPO, 2003).

6. See American Library Association Web site on the USA Patriot Act at www.ala.org/Content/NavigationMenu/Our_Association/Offices/Intellectual_Freedom3/Intellectual_Freedom_Issues/USA_Patriot_Act.htm. Also see Leigh S. Estabrook, *Public Libraries and Civil Liberties: A Profession Divided* (Urbana, Ill.: Library Research Center, 2002). Available at http://alexia.lis.uiuc.edu/gslis/research/civil_liberties.html.

7. American Library Association, *Guideline for Librarians on the USA Patriot Act* (Washington, D.C.: American Library Association, 2002). Also see Eric Lichtblau, "Justice Dept. Lists Use of New Power to Fight Terror," *New York Times,* May 21, 2003.

8. See Wayne Wiegand, "To Reposition a Research Agenda: What American Studies Can Teach the LIS Community About the Library in the Life of the User," *Library Quarterly,* January 2004.

9. C. Wright Mills, *The Power Elite* (New York: Oxford University Press, 1956), p. 3.

10. Amartya Kumar Sen, *Development as Freedom* (New York: Knopf, 1999). Also see Joseph E. Stiglitz, *Information and Capital Markets,* Working Paper/National Bureau of Economic Research, no. 678 (Cambridge, Mass.: National Bureau of Economic Research, 1981); Stiglitz, *Globalization and Its Discontents* (New York: Norton, 2002).

11. See C. Wright Mills, *The Power Elite* (New York: Oxford University Press, 2000). Also see Jim Miller, *Democracy Is in the Streets* (New York: Simon & Schuster, 1987).

12. Eli Noam, lecture, Department of Culture and Communication, New York University, October 29, 2002.

Chapter 9

1. Elena Bonner, "The Remains of Totalitarianism," *New York Review of Books,* March 8, 2001. Also see Siva Vaidhyanathan, "Fired for Being Israeli," Salon.com, June 26, 2002.

2. Robert Merton, "Science and Democratic Social Structure," in *Social Theory and Social Structure* (New York: Free Press, 1968), pp. 604–615. Also see Freeman Dyson, "In Praise of Amateurs," *New York Review of Books,* December 5, 2002; Timothy Ferris, "The Gentleman Scientist," *New York Review of Books,* March 27, 2003.

3. Will Knight, "Computer Scientists Boycott US over Digital Copyright Law," NewScientist.com, July 23, 2001.

4. Associated Press, "Researchers Stymied by Block on Government Documents," Siliconvalley.com, October 14, 2002.

5. Peg Brickley, "New Antiterrorism Tenets Trouble Scientists," *Scientist,* October 28, 2002.

6. Roger Tatoud, "Copyright and Science: Gridlocking Knowledge?" openDemocracy.net, June 24, 2002.

7. See Budapest Open Access Initiative Web site at www.soros.org/openaccess. Also see John Willinsky, "Copyright Contradictions in Scholarly Publishing," *First Monday,* November 2002. Available at http://firstmonday.org/issues/issue7_11/willinsky/index.html. Also see Nicholas Thompson, "May the Source Be with You: Can a Band of Biologists Who Share Data Freely Out-Innovate the Corporate Researchers Who Hoard It?" *Washington Monthly,* July–August 2002. Also see Amy

Harmon, "New Premise in Science: Get the Word Out Quickly, Online," *New York Times*, December 17, 2002.

8. *Diamond v. Chakrabarty*, 447 U.S. 303 (1980).

9. John Sulston, "Heritage of Humanity," *Le Monde Diplomatique*, December 2002. Also see John Sulston and Georgina Ferry, *The Common Thread: A Story of Science, Politics, Ethics, and the Human Genome* (Washington, D.C.: Joseph Henry, 2002).

10. Merck, the giant American pharmaceutical company, worked throughout the 1990s to put information about expressed sequence tags, or ESTs, into the public domain. See Eliot Marshall, "A Deluge of Patents Creates Legal Hassles for Researchers," *Science*, April 14, 2000. Also see Arti K. Rai, "Fostering Cumulative Innovation in the Biopharmaceutical Industry: The Role of Patents and Antitrust," *Berkeley Technology Law Journal*, Spring 2001; Rai, "Regulating Scientific Research: Intellectual Property Rights and the Norms of Science," *Northwestern University Law Review*, Fall 1999; Michael Heller and Rebecca Eisenberg, "Can Patents Deter Innovation? The Anticommons in Biomedical Research," *Science*, May 1, 1998.

11. Arti K. Rai and Rebecca Eisenberg, "Bayh-Dole Reform and the Progress of Biomedicine," *Law and Contemporary Problems*, Winter Spring 2003.

12. Sulston, *Common Thread*, pp. 228–229.

13. Werner R. Loewenstein, *The Touchstone of Life: Molecular Information, Cell Communication, and the Foundations of Life* (New York: Oxford University Press, 1999), pp. 6–10. I was inspired by a book called *Regulation and Control Mechanisms in Biological Systems* written by a brilliant scientist who teaches at SUNY-Buffalo. He not only taught me just about everything I needed to write this chapter, he gave me half my genes. His name is Dr. Vishnampet Sivaramakrishnan Vaidhyanathan. Also, I blame him for my impending baldness.

14. Tim Beardsley, "Piecemeal Patents: DNA Gene Labeling," *Scientific American*, July 1992, p. 106.

15. Horace Freeland Judson, "A History of the Science and Technology Behind Gene Mapping and Sequencing," in Daniel J. Kevles and Leroy E. Hood, eds., *The Code of Codes: Scientific and Social Issues in the Human Genome Project* (Cambridge, Mass.: Harvard University Press, 1992).

16. "The End of the Beginning," *Economist*, October 24, 1992, p. 96.

17. Christine Gorman, "The Race to Map Our Genes," *Time*, February 8, 1993, p. 57.

18. "End of the Beginning," p. 98.

19. Christopher Anderson, "NIH cDNA Patent Rejected; Backers Want to Amend Law," *Nature*, September 24, 1992, p. 263.

20. "Mapping Mankind," p. 16.

21. Sulston, *Common Thread*, pp. 211–212.

22. For a summary of copyright law and its growing lack of enforceability, see Peter Jaszi, "On the Author Effect: Contemporary Copyright and Collective Creativity," *Arts and Entertainment Law Journal* 10, no. 2 (1992).

23. Sabra Chartbrand, "In Health Emergencies, Brazil Allows the Copying of Drugs, to the Dismay of American Companies," *New York Times*, February 19, 2001, p. C8.

24. Thomas Cech, "Working Together in the Biology Revolution," *Chronicle of Higher Education*, February 16, 2001, p. B24.

25. Stephen Jay Gould, "Humbled by the Genome's Mysteries," *New York Times*, February 19, 2001, p. A15.

26. David Malakoff, "Will a Smaller Genome Complicate the Patent Chase?" *Science*, February 16, 2001, p. 1194.

27. Evelyn Fox Keller, *The Century of the Gene* (Cambridge, Mass.: Harvard University Press, 2000).

28. J. Cello, A. V. Paul, and E. Wimmer, "Chemical Synthesis of Poliovirus Cdna: Generation of Infectious Virus in the Absence of Natural Template," *Science* 297, no. 5583 (2002); Cello, Paul, and Wimmer, "Vaccines Should Be Kept Even If Polio Is Wiped Out," *Nature* 418, no. 6901 (2002).

29. Amy Harmon, "Journal Editors to Consider U.S. Security in Publishing," *New York Times*, February 16, 2003, p. 15.

30. Sue Anne Presley and Justin Blum, "Hijackers May Have Accessed Computers at Public Libraries," *Washington Post*, September 17, 2001.

31. Siva Vaidhyanthan, "Putting a Long on E-Books: A New Cold War Looms Over Your Right to Read," MSNBC.COM, July 19, 2001. Available at http://stacks.msnbc.com/news/602444.asp.

32. Andrew Light, *Reel Arguments: Film, Philosophy, and Social Criticism* (Boulder: Westview, 2003).

33. "What Is Freenet?" Available at http://freenetproject.org/cgi-bin/twiki/view/Main/WhatIs.

34. Benkler, "Coase's Penguin." The operative protocol of the free software movement is under overt attack from the commercial software world, primarily Microsoft. Microsoft officials have been warning that use and adoption of free software and its powerful "negative copyright" feature, the general public license, threatens the entire information economy by wrenching property into the public domain. This is not just an exaggeration; it is a lie. Regardless of the ethics of the participants in the battle between free and proprietary software, powerful interests consider free software threatening because it is open, free, and distributed—and often better.

35. Eric G. Campbell et al., "Data Withholding in Academic Genetics," *Journal of the American Medical Association*, January 23, 2002.

Chapter 10

1. Robert Wright, *Nonzero: The Logic of Human Destiny* (New York: Vintage, 2001). Also see Thomas L. Friedman, *The Lexus and the Olive Tree* (New York: An-

chor, 2000); John Micklethwait and Adrian Wooldridge, *A Future Perfect: The Challenge and Hidden Promise of Globalization* (New York: Times Books, 2000). Also see Micklethwait and Wooldridge, "Rebuilding the Alliance to Rebuild Globalization," *New York Times,* April 13, 2003.

2. Christopher May, *The Information Society: A Sceptical View* (Cambridge: Polity, 2002).

3. See Joseph E. Stiglitz, *Globalization and Its Discontents* (New York: Norton, 2002).

4. Richard N. Rosecrance, *The Rise of the Virtual State: Wealth and Power in the Coming Century* (New York: Basic, 1999). Also see Friedman, *The Lexus and the Olive Tree;* Carlos María Correa, *Intellectual Property Rights, the WTO, and Developing Countries: The Trips Agreement and Policy Options* (London: Zed, 2000). Also see Hans-Henrik Holm and Georg Sorensen, *Whose World Order? Uneven Globalization and the End of the Cold War* (Boulder, Colo.: Westview, 1995).

5. This is known as the "moral hazard" of economic stabilization guarantees. See Stiglitz, *Globalization and Its Discontents.*

6. This case of Brazil trying to develop a national passenger aircraft industry may be hypothetical, but the issues are real. Brazil does have a domestic military aircraft industry. When neighboring Colombia sought to purchase light fighter aircraft from Brazil in 2002, General James Hill, commander of the U.S. Southern Command, warned the Colombian government that it should buy U.S.-made C-130 aircraft. If Colombia continued its plans to purchase Brazilian planes, the general wrote, it would jeopardize future American aid. See William Finnegan, "The Economics of Empire: Notes on the Washington Consensus," *Harper's,* May 2003.

7. Naomi Klein and Debra Ann Levy, *Fences and Windows: Dispatches from the Front Lines of the Globalization Debate* (New York: Picador USA, 2002). Also see Naomi Klein, *No Space, No Choice, No Jobs, No Logo: Taking Aim at the Brand Bullies* (New York: Picador USA, 2000).

8. May, Information Society. Also see Richard Barbrook, "The California Ideology," in Peter Ludlow, *Crypto Anarchy, Cyberstates, and Pirate Utopias,* Digital Communication (Cambridge, Mass.: MIT Press, 2001). Also see Barbrook, "The Gift of the Net." Available at www.opendemocracy.net/debates/article-8-101-1468.jsp. Lawrence Lessig has described the conflicts between Northern and Southern California as "a house divided." Northern California—home of Silicon Valley—"believes in a free exchange of ideas." Hollywood desires control over its copyrighted material and sees it as "their plantation and seek permission from the master. . . . If you develop technology that interferes with their right of perfect control, you will be punished." See David Sims, "Lessig: North v. South," *O'Reilly Network,* August 29, 2001. Available at www.oreillynet.com/lpt/wlg/635. This "civil war" cooled in 2003, as computer and software companies recognized markets in digital rights management that they could exploit with the help of Hollywood.

9. May, *Information Society*. Also see Daniel H. Pink, *Free Agent Nation: How America's New Independent Workers Are Transforming the Way We Live* (New York: Warner Books, 2001); Richard L. Florida, *The Rise of the Creative Class: And How It's Transforming Work, Leisure, Community, and Everyday Life* (New York: Basic, 2002).

10. George F. Gilder, *Telecosm: How Infinite Bandwidth Will Revolutionize Our World* (New York: Free Press, 2000). Also see George F. Gilder, *The Spirit of Enterprise* (New York: Simon & Schuster, 1984). For a scathing criticism of Gilder, see Thomas Frank, *One Market Under God: Extreme Capitalism, Market Populism, and the End of Economic Democracy* (New York: Doubleday, 2000). For Drucker's vision, see Peter Ferdinand Drucker, *The Age of Discontinuity: Guidelines to Our Changing Society* (New Brunswick, N.J.: Transaction, 1992). Also see Peter Ferdinand Drucker, *Post-Capitalist Society* (New York: HarperBusiness, 1993).

11. Kevin Kelly, *New Rules for the New Economy: 10 Radical Strategies for a Connected World* (New York: Viking, 1998). Kelly instructed business leaders how to thrive under the new self-justifying "rules," while I am warning citizens to pay close attention to the assumptions at work in the promulgation of these "rules." Also see Kevin Kelly, *Out of Control: The Rise of Neo-Biological Civilization* (Reading, Mass.: Addison-Wesley, 1994). For a stinging critique of Kelly and his effects on economic "thinking" in the United States, see Frank, *One Market Under God*.

12. Edward Luttwak, *Turbo-Capitalism: Winners and Losers in the Global Economy* (New York: HarperCollins, 1999).

13. Andrew Leonard, "Enron, We Hardly Knew Ye," Salon.com, November 9, 2001. Available at http://archive.salon.com/tech/col/leon/2001/11/09/enron/print.html. Also see Robert Sheer, "Enron and the End of the Reagan Revolution," Salon.com, February 13, 2002. Available at www.salon.com/news/col/scheer/2002/02/13/enron/index.html.

14. Rosecrance, *Rise of the Virtual State*.

15. Eddie Yuen, Daniel Burton-Rose, and George N. Katsiaficas, *The Battle of Seattle: The New Challenge to Capitalist Globalization* (New York: Soft Skull, 2002). Also see Klein and Levy, *Fences and Windows*.

16. David F. Ronfeldt, John Arquilla, Graham E. Fuller, Melissa Fuller, *The Zapatista "Social Netwar" in Mexico* (Washington, D.C.: RAND Corporation, 1998). Also see "Zapatistas in Cyberspace." Available at www.eco.utexas.edu/faculty/Cleaver/zapsincyber.html.

17. David Graeber, "The New Anarchists," *New Left Review*, January—February 2002.

18. Graeber, "New Anarchists."

19. Nathan Newman, "Who Killed Carlo Giuliani?" Available at www.globalpolicy.org/ngos/role/globdem/globprot/2001/0802kill.htm. Also see "G8 Summit Death Shocks Leaders," CNN.com, July 21, 2001. Available at www.cnn.com/2001/WORLD/europe/07/20/genoa.protests. Also see Ramor Ryan, "Death and Terror in Genoa." Available at http://struggle.ws/global/genoa/ramor.html.

Chapter 11

1. Available at www.whitehouse.gov/nsc/nss.html.

2. Noam Chomsky, "The Crimes of Intcom," *Foreign Policy,* September–October 2002, pp. 34–35.

3. Kofi Annan, "Problems Without Passports," *Foreign Policy,* September–October 2002, pp. 30–31.

4. Annan, "Problems Without Passports." For an indictment of the representative nature of the World Bank and the International Monetary Fund, see Joseph Stiglitz, *Globalization and Its Discontents* (New York: Norton, 2002).

5. Arjun Appadurai, "Broken Promises," *Foreign Affairs,* September–October 2002, pp. 42–43.

6. Marc Cooper, "From Protest to Politics," *Nation,* March 14, 2002. Available at www.alternet.org/story.html?StoryID=12624. See Naomi Klein, "Porto Alegre, Brazil: 'Bad Capitalist! No Martini.' Do The Public Floggings at the World Economic Forum Represent True Progress?" Available at www.tompaine.com/feature.cfm/ ID/5103. Also see Jennifer Block, "Today, Porto Alegre. Tomorrow . . . ?" *Mother Jones,* February 6, 2002. Available at www.motherjones.com/web_exclusives/ features/news/world_social_forum.html. Also see Walden Bello, "Battling Barbarism," *Foreign Policy,* September–October 2002, pp. 41–42.

7. Monroe Edwin Price, *Media and Sovereignty: The Global Information Revolution and Its Challenge to State Power* (Cambridge, Mass.: MIT Press, 2002).

8. "The National Security Strategy of the United States of America." Available at www.whitehouse.gov/nsc/nss.html. September 2002, p. 1.

9. "National Security Strategy of the United States of America," p. 1.

10. "National Security Strategy of the United States of America," p. 1.

11. John Arquilla and David F. Ronfeldt, *Networks and Netwars: The Future of Terror, Crime, and Militancy* (Santa Monica, Calif.: Rand, 2001).

12. See Alessandro Politi, "The Citizen as Intelligence Minuteman," *International Journal of Intelligence and Counterintelligence* 16 (2003).

13. Peter Bergen, "Al Qaeda's New Tactics," *New York Times,* November 15, 2002, p. A27. Also see Peter L. Bergen, *Holy War, Inc.: Inside the Secret World of Osama bin Laden* (New York: Free Press, 2001).

14. Bergen, *Holy War,* p. 200.

15. Gilles Kepel, *Jihad: The Trail of Political Islam* (Cambridge, Mass.: Belknap Press of Harvard University Press, 2002). Also see John L. Esposito, *Islam: The Straight Path,* 3d ed. (New York: Oxford University Press, 1998); Esposito, *The Oxford History of Islam* (Oxford: Oxford University Press, 1999); Esposito, *Unholy War: Terror in the Name of Islam* (Oxford: Oxford University Press, 2002); John L. Esposito and François Burgat, *Modernizing Islam: Religion in the Public Sphere in the Middle East and Europe* (New Brunswick, N.J.: Rutgers University Press, 2003).

16. Hannah Arendt and P. R. Baehr, *The Portable Hannah Arendt* (New York: Penguin, 2000).

17. Jeffrey Rosen, "Silicon Valley's Spy Game," *New York Times Magazine,* April 14, 2002, p. 46.

18. Steve Lohr, "Data Expert Is Cautious About Misuse of Information," *New York Times,* March 25, 2003, p. C6.

19. Lohr, "Data Expert." Also see www.the-data-mine.com.

20. "AI NEWS FOCUS May 1999 People's Republic of China: Ten years after Tiananmen . . . and Still Waiting for Justice," www.amnesty.org/ailib/intcam/china/focus2.htm. Also see Jonathan Mirsky, "Nothing to Celebrate: China's Wasted Half Century," *New Republic,* October 11, 1999. Available at www.thenewrepublic.com/current/coverstory101199.html.

21. Tina Rosenberg, "John Kamm's Third Way," *New York Times Magazine,* March 3, 2002.

22. Ian Buruma, *Bad Elements: Chinese Rebels from Los Angeles to Beijing* (New York: Random House, 2001).

23. Quoted in Clifford Geertz, *Available Light: Anthropological Reflections on Philosophical Topics* (Princeton, N.J.: Princeton University Press, 2000), 233. Also see Benedict R. O'G Anderson, *Imagined Communities: Reflections on the Origin and Spread of Nationalism,* rev. and ext. ed. (London: Verso, 1991).

24. Ian Buruma, "China in Cyberspace," *New York Review of Books,* November 4, 1999.

25. Associated Press, "China Shuts 17,000 Internet Bars," November 21, 2001.

26. Buruma, *Bad Elements.* Also see Danny Schechter, *Falun Gong's Challenge to China: Spiritual Practice or "Evil Cult"? A Report and Reader* (New York: Akashic, 2000).

27. Shanthi Kalathil and Taylor C. Boas, *Open Networks, Closed Regimes: The Impact of the Internet on Authoritarian Regimes* (Washington, D.C.: Carnegie Endowment for International Peace, 2003). Also see Buruma, *Bad Elements.*

28. Chris Sprigman, "Hacking for Free Speech." Available at http://practice.findlaw.com/hack-0703.html. Also see www.hacktivismo.com.

29. Jennifer Lee, "Guerilla Warfare, Waged with Code," *New York Times,* October 10, 2002. Also see Andrew Oram, ed., *Peer-to-Peer: Harnessing the Power of Disruptive Technologies* (Cambridge, Mass.: O'Reilly, 2001); http://freenet.sourceforge.net.

30. Robert D. Kaplan, *The Coming Anarchy: Shattering the Dreams of the Post Cold War* (New York: Random House, 2000). Originally put forth as an article: Robert D. Kaplan, "The Coming Anarchy," *Atlantic Monthly,* February 1994.

31. Jennifer Widner, "States and Statelessness in Late Twentieth-Century Africa," *Daedalus,* 1995, 148.

32. Widner, "States and Statelessness." Also see Mark Granovetter, "Economic Action and Social Structure: The Problem of Embeddedness," *American Journal of*

Sociology 91 (1985); James Coleman, *Foundations of Social Theory* (Cambridge, Mass.: Harvard University Press, 1990).

33. Howard Rheingold, *Smart Mobs: The Next Social Revolution* (Cambridge, Mass.: Perseus, 2002).

Chapter 12

1. Walter Lippmann, *Public Opinion* (New York: Harcourt Brace, 1922).

2. Eric Schmidt, "Rumsfeld Says He May Drop New Office of Influence," *New York Times*, February 25, 2002, p. A13.

3. Siva Vaidhyanathan, *Copyrights and Copywrongs: The Rise of Intellectual Property and How It Threatens Creativity* (New York: New York University Press, 2001).

4. Herbert J. Gans, *Popular Culture and High Culture: An Analysis and Evaluation of Taste*, rev. and updated ed. (New York: Basic, 1999). Also see Erika Lee Doss, *Spirit Poles and Flying Pigs: Public Art and Cultural Democracy in American Communities* (Washington, D.C.: Smithsonian Institution Press, 1995); David Trend, *Cultural Democracy: Politics, Media, New Technology*, SUNY Series, Interruptions: Border Testimony(Ies) and Critical Discourse/S (Albany: State University of New York Press, 1997); Lambert Zuidervaart and Henry Luttikhuizen, *The Arts, Community, and Cultural Democracy: Cross-Currents in Religion and Culture* (New York: St. Martin's, 2000).

5. Iseult Honohan, *Civic Republicanism*, Problems of Philosophy Series (New York: Routledge, 2002); Adrian Oldfield, *Citizenship and Community: Civic Republicanism and the Modern World* (London: Routledge, 1990). Also see Gisela Bock, Quentin Skinner, and Maurizio Viroli, *Machiavelli and Republicanism: Ideas in Context* (Cambridge: Cambridge University Press, 1990); John R. Howe, *The Role of Ideology in the American Revolution*, American Problem Studies (New York: Holt, Rinehart & Winston, 1970); Rogers M. Smith, *Civic Ideals: Conflicting Visions of Citizenship in U.S. History*, Yale ISPS Series (New Haven, Conn.: Yale University Press, 1997); Maurizio Viroli, *Republicanism* (New York: Hill & Wang, 2002).

BIBLIOGRAPHY

Alderman, John. *Sonic Boom: Napster, Mp3, and the New Pioneers of Music*. Cambridge, Mass.: Perseus, 2001.

American Library Association. *Guideline for Librarians on the USA Patriot Act*. Washington, D.C.: American Library Association, 2002.

Anderson, Benedict R. O'G. *Imagined Communities: Reflections on the Origin and Spread of Nationalism*. Rev. and extended ed. London: Verso, 1991.

_____. *The Spectre of Comparisons: Nationalism, Southeast Asia, and the World*. London: Verso, 1998.

Anderson, Benedict R. O'G, and Gail Kligman. *Long-Distance Nationalism: World Capitalism and the Rise of Identity Politics*. Center for German and European Studies, Working Paper no. 5.1. Berkeley: Center for German and European Studies, University of California, 1992.

Andrews, William L., Frances Smith Foster, and Trudier Harris. *The Oxford Companion to African American Literature*. New York: Oxford University Press, 1997.

Annas, George J., and Sherman Elias. *Gene Mapping: Using Law and Ethics as Guides*. New York: Oxford University Press, 1992.

Appadurai, Arjun. *Globalization: A Millennial Quartet Book*. Durham, N.C.: Duke University Press, 2001.

_____. *Modernity at Large: Cultural Dimensions of Globalization*. Vol. 1 of *Public Worlds*. Minneapolis: University of Minnesota Press, 1996.

_____. *The Social Life of Things: Commodities in Cultural Perspective*. Cambridge: Cambridge University Press, 1986.

Arendt, Hannah. *The Human Condition*. 2d ed. Chicago: University of Chicago Press, 1998.

_____. *The Origins of Totalitarianism*. New York: Harcourt Brace & World, 1968.

Arendt, Hannah, and P. R. Baehr. *The Portable Hannah Arendt*. Viking Portable Library. New York: Penguin, 2000.

Arnold, Matthew, and Stefan Collini. *Culture and Anarchy and Other Writings*. Cambridge Texts in the History of Political Thought. Cambridge: Cambridge University Press, 1993.

Arnott, Richard J., Bruce C. N. Greenwald, and Joseph E. Stiglitz. *Information and Economic Efficiency*. NBER Working Paper Series, no. 4533. Cambridge, Mass.: National Bureau of Economic Research, 1993.

Arnott, Richard, and Joseph E. Stiglitz. *Randomization with Asymmetric Information*. NBER Working Paper, no. 2507. Cambridge, Mass.: National Bureau of Economic Research, 1988.

Arquilla, John, and David F. Ronfeldt. *The Advent of Netwar*. Santa Monica, Calif.: Rand, 1996.

———. *The Emergence of Neopolitik: Toward an American Information Strategy*. Santa Monica, Calif.: Rand, 1999.

———. *In Athena's Camp: Preparing for Conflict in the Information Age*. Santa Monica, Calif.: Rand, 1997.

———. *Networks and Netwars: The Future of Terror, Crime, and Militancy*. Santa Monica, Calif.: Rand, 2001.

———. *Swarming and the Future of Conflict*. DB-311-OSD. Santa Monica, Calif.: Rand, 2000.

Aufderheide, Patricia. *Communications Policy and the Public Interest*. Guilford Communication Series. New York: Guilford, 1999.

———. *Conglomerates and the Media*. New York: New Press/Norton, 1997.

———. *The Daily Planet: A Critic on the Capitalist Culture Beat*. Minneapolis: University of Minnesota Press, 2000.

Aunger, Robert. *The Electric Meme: A New Theory of How We Think*. New York: Free Press, 2002.

Avrich, Paul. *The Haymarket Tragedy*. Princeton, N.J.: Princeton University Press, 1984.

Aylmer, G. E. *The Levellers in the English Revolution*. Documents of Revolution. Ithaca, N.Y.: Cornell University Press, 1975.

Bailyn, Bernard. *The Ideological Origins of the American Revolution*. Cambridge, Mass.: Belknap Press of Harvard University Press, 1967.

Bakunin, Mikhail Aleksandrovich, and Sam Dolgoff. *Bakunin on Anarchy*. New York: Knopf, 1972.

Balkin, J. M. *Cultural Software: A Theory of Ideology*. New Haven, Conn.: Yale University Press, 1998.

Barabási, Albert-Laszló. *Linked: The New Science of Networks*. Cambridge, Mass.: Perseus, 2002.

Barber, Benjamin R. *Superman and Common Men: Freedom, Anarchy, and the Revolution*. New York: Praeger, 1971.

Becker, Lawrence C. *A New Stoicism*. Princeton, N.J.: Princeton University Press, 1998.

Bergen, Peter L. *Holy War, Inc.: Inside the Secret World of Osama bin Laden*. New York: Free Press, 2001.

Berger, Arthur Asa. *The Postmodern Presence: Readings on Postmodernism in American Culture and Society*. Walnut Creek, Calif.: AltaMira, 1998.

Berkowitz, Bruce. *The New Face of War: How War Will Be Fought in the Twenty-first Century*. New York: Free Press, 2003.

Berlin, Isaiah. *The Age of Enlightenment: The Eighteenth-Century Philosophers*. New York: New American Library, 1984.

_____. *The Hedgehog and the Fox: An Essay on Tolstoy's View of History*. New York: Simon & Schuster, 1986.

Berlin, Isaiah, and Henry Hardy. *The Crooked Timber of Humanity: Chapters in the History of Ideas*. Princeton, N.J.: Princeton University Press, 1997.

_____. *The Sense of Reality: Studies in Ideas and Their History*. New York: Farrar Straus & Giroux, 1997.

_____. *Three Critics of the Enlightenment: Vico, Hamann, Herder*. Princeton, N.J.: Princeton University Press, 2000.

Berlin, Isaiah, Henry Hardy, and Roger Hausheer. *The Proper Study of Mankind: An Anthology of Essays*. New York: Farrar Straus & Giroux, 2000.

Berlinski, David. *The Advent of the Algorithm: The Idea That Rules the World*. New York: Harcourt, 2000.

Berman, Paul. *Quotations from the Anarchists*. New York: Praeger, 1972.

_____. *Terror and Liberalism*. New York: Norton, 2003.

Berners-Lee, Tim, and Mark Fischetti. *Weaving the Web: The Original Design and Ultimate Destiny of the World Wide Web by Its Inventor*. New York: HarperCollins, 2000.

Bewes, Timothy. *Cynicism and Postmodernity*. London: Verso, 1997.

Biegel, Stuart. *Beyond Our Control? Confronting the Limits of Our Legal System in the Age of Cyberspace*. Cambridge, Mass.: MIT Press, 2001.

Bock, Gisela, Quentin Skinner, and Maurizio Viroli. *Machiavelli and Republicanism*. Ideas in Context. Cambridge: Cambridge University Press, 1990.

Bollier, David. *Silent Theft: The Private Plunder of Our Common Wealth*. New York: Routledge, 2002.

Bonner, Elena. "The Remains of Totalitarianism." *New York Review of Books*, March 8, 2001.

Borgmann, Albert. *Crossing the Postmodern Divide*. Chicago: University of Chicago Press, 1992.

_____. *Holding On to Reality: The Nature of Information at the Turn of the Millennium*. Chicago: University of Chicago Press, 1999.

_____. *The Philosophy of Language: Historical Foundations and Contemporary Issues*. The Hague: Nijhoff, 1974.

_____. *Technology and the Character of Contemporary Life: A Philosophical Inquiry*. Chicago: University of Chicago Press, 1984.

Borsook, Paulina. *Cyberselfish: A Critical Romp Through the Terribly Libertarian Culture of High Tech*. New York: PublicAffairs, 2000.

Bourdieu, Pierre. *Distinction: A Social Critique of the Judgment of Taste*. Cambridge, Mass.: Harvard University Press, 1984.

———. *In Other Words: Essays Toward a Reflexive Sociology*. Oxford: Polity, 1990.

———. *The Logic of Practice*. Cambridge: Polity, 1990.

———. *Practical Reason: On the Theory of Action*. Stanford, Calif.: Stanford University Press, 1998.

Bourdieu, Pierre, and Randal Johnson. *The Field of Cultural Production: Essays on Art and Literature*. New York: Columbia University Press, 1993.

Bourdieu, Pierre, and Jean Claude Passeron. *Reproduction in Education, Society, and Culture*. London: Sage/Theory Culture and Society Department of Administrative and Social Studies Teesside Polytechnic, 1990.

Boyle, James. *Critical Legal Studies: The International Library of Essays in Law and Legal Theory*. New York: New York University Press, 1992.

———. *Shamans, Software, and Spleens: Law and the Construction of the Information Society*. Cambridge, Mass.: Harvard University Press, 1997.

Bradford, Gigi, Michael Gary, and Glenn Wallach. *The Politics of Culture: Policy Perspectives for Individuals, Institutions, and Communities*. New York: New Press/Norton, 2000.

Branham, Robert Bracht, and Marie-Odile Goulet-Cazé. *The Cynics: The Cynic Movement in Antiquity and Its Legacy*. Hellenistic Culture and Society, no. 23. Berkeley: University of California Press, 1996.

Branscomb, Lewis M., Fumio Kodama, and Richard L. Florida. *Industrializing Knowledge: University-Industry Linkages in Japan and the United States*. Cambridge, Mass.: MIT Press, 1999.

Breckenridge, Carol Appadurai. *Cosmopolitanism*. A Millennial Quartet Book. Durham, N.C.: Duke University Press, 2002.

Buchanan, Mark. *Nexus: Small Worlds and the Groundbreaking Science of Networks*. New York: Norton, 2002.

Buruma, Ian. *Bad Elements: Chinese Rebels from Los Angeles to Beijing*. New York: Random House, 2001.

———. *The Missionary and the Libertine: Love and War in East and West*. London: Faber & Faber, 1996.

Calhoun, Craig J. *Social Theory and the Politics of Identity*. Oxford: Blackwell, 1994.

Carey, James W. *Communication as Culture: Essays on Media and Society, Media, and Popular Culture*. Vol. 1. New York: Routledge, 1992.

Castells, Manuel. "The Power of Identity." In *Information Age: Economy, Society, and Culture*. Vol. 2. Malden, Mass.: Blackwell, 1997.

———. *The Rise of the Network Society*. Edited by Manuel Castells. Vol. 1 of *Information Age*. 2d ed. Oxford: Blackwell, 2000.

Cello, J., A. V. Paul, and E. Wimmer. "Chemical Synthesis of Poliovirus Cdna: Generation of Infectious Virus in the Absence of Natural Template." *Science* 297, no. 5583 (2002): 1016–1018.

———. "Vaccines Should Be Kept Even If Polio Is Wiped Out." *Nature* 418, no. 6901 (2002): 915.

Centre for the Study of Developing Societies. *The Public Domain*. Sarai Reader, no. 1. Delhi, Amsterdam: Society for Old and New Media, 2001.

Centre for the Study of Developing Societies and Society for Old and New Media. *The Cities of Everyday Life*. Sarai Reader, no. 2. New Delhi: Centre for the Study of Developing Societies/Society for Old and New Media/Rainbow, 2002.

Chaloupka, William. *Everybody Knows: Cynicism in America*. Minneapolis: University of Minnesota Press, 1999.

Chang, Ha-Joo, ed. *Joseph Stiglitz and the World Bank: The Rebel Within*. London: Anthem, 2001.

Chang, Nancy. *Silencing Political Dissent*. New York: Seven Stories, 2002.

Cheung, David Graham J. Williams, and Qing Li. *Proceedings of the Fifth Pacific-Asia Conference on Knowledge Discovery and Data Mining (PAKDD 2001)*. Lecture Notes in Artificial Intelligence. Berlin: Springer, 2001.

Chisholm, Donald William. *Coordination Without Hierarchy: Informal Structures in Multiorganizational Systems*. Berkeley: University of California Press, 1989.

Cohen, Randy. *The Good, the Bad, and the Difference: How to Tell Right from Wrong in Everyday Situations*. New York: Doubleday, 2002.

Colebrook, Claire. *Gilles Deleuze*. Routledge Critical Thinkers. London: Routledge, 2002.

Coleman, James. *Foundations of Social Theory*. Cambridge, Mass.: Harvard University Press, 1990.

Colish, Marcia L. *The Stoic Tradition from Antiquity to the Early Middle Ages*. Leiden, Neth.: Brill, 1990.

Conway, Maura. "Reality Bytes: Cyberterrorism and the Terrorist 'Use' of the Internet." *First Monday* 7, no. 11 (2002).

Coram, Alex Talbot. *State, Anarchy, and Collective Decisions*. New York: Palgrave, 2001.

Correa, Carlos María. *Intellectual Property Rights, the WTO, and Developing Countries: The Trips Agreement and Policy Options*. London: Zed, 2000.

———. "The Management of International Intellectual Property." *International Journal of Technology Management* 10, no. 2–3 (1995).

———. *Regímenes de control de la transferencia de tecnología en América latina*. Serie Monografías/Banco Interamericano de Desarrollo Instituto para la Integración de America Latina, no. 5. Buenos Aires: Instituto para la Integración de América Latina Banco Interamericano de Desarrollo, 1979.

Correa, Carlos María, and Abdulqawi Yusuf. *Intellectual Property and International Trade: The TRIPS Agreement*. London: Kluwer Law International, 1998.

Cowen, Tyler. *Creative Destruction: How Globalization Is Changing the World's Cultures*. Princeton, N.J.: Princeton University Press, 2002.

———. *In Praise of Commercial Culture*. Cambridge, Mass.: Harvard University Press, 1998.

Crews, Kenneth D. *Copyright Essentials for Librarians and Educators*. Chicago: American Library Association, 2000.

_____. *Copyright, Fair Use, and the Challenge for Universities: Promoting the Progress of Higher Education*. Chicago: University of Chicago Press, 1993.

_____. *University Copyright Policies in ARL Libraries*. Spec Kit no. 138. Washington, D.C.: Association of Research Libraries Office of Management Studies, 1987.

Darnton, Robert. *The Business of Enlightenment: A Publishing History of the Encyclopédie, 1775–1800*. Cambridge, Mass.: Belknap Press of Harvard University Press, 1979.

_____. *The Forbidden Best-Sellers of Prerevolutionary France*. New York: Norton, 1995.

_____. *The Great Cat Massacre and Other Episodes in French Cultural History*. New York: Basic, 1984.

_____. *The Literary Underground of the Old Regime*. Cambridge, Mass.: Harvard University Press, 1982.

_____. "No Computer Can Hold the Past." *New York Times*, June 12, 1999.

_____. *What Was Revolutionary About the French Revolution?* Charles Edmonson Historical Lectures, no. 11. Waco, Tex.: Baylor University Press/Markham Press Fund, 1990.

Dees, Morris, and James Corcoran. *Gathering Storm: America's Militia Threat*. New York: HarperCollins, 1996.

Deleuze, Gilles. *Negotiations, 1972–1990: European Perspectives*. New York: Columbia University Press, 1995.

Dewey, John. *The Public and Its Problems*. New York: Holt, 1927.

Dias, Reginald Walter Michael, and Graham Hughes. *Jurisprudence*. London: Butterworth, 1957.

Diogenes, Laertius. *Lives of Eminent Philosophers, with an English Translation*. Translated by Robert Drew Hicks. Cambridge, Mass.: Harvard University Press, 1942.

Doss, Erika Lee. *Spirit Poles and Flying Pigs: Public Art and Cultural Democracy in American Communities*. Washington, D.C.: Smithsonian Institution Press, 1995.

Drabble, Margaret. *The Oxford Companion to English Literature*. 6th ed. Oxford: Oxford University Press, 2000.

Drahos, Peter. *Intellectual Property*. International Library of Essays in Law and Legal Theory. 2d series. Aldershot, U.K.: Ashgate/Dartmouth, 1999.

_____. *A Philosophy of Intellectual Property*. Aldershot, U.K.: Dartmouth, 1996.

Drahos, Peter, and Ruth Mayne. *Global Intellectual Property Rights: Knowledge, Access, and Development*. New York: Palgrave, 2002.

Drucker, Peter Ferdinand. *The Age of Discontinuity: Guidelines to Our Changing Society*. New Brunswick, N.J.: Transaction, 1992.

_____. *Post-Capitalist Society*. New York: HarperBusiness, 1993.

Druschel, Peter, Frans Kaashoek, and Antony Rowstron. *Peer-to-Peer Systems: First International Workshop (IPTPS 2002)*. Lecture Notes in Artificial Intelligence. Berlin: Springer, 2002.

Dyson, Esther. *Release 2.1: A Design for Living in the Digital Age.* New York: Broadway, 1998.

Dyson, Freeman J. *The Sun, the Genome, and the Internet: Tools of Scientific Revolutions.* New York: New York Public Library/Oxford University Press, 1999.

Dyson, Kenneth H. F., and Walter Homolka. *Culture First: Promoting Standards in the New Media Age.* London: Cassell, 1996.

Eagleton, Terry. *Ideology: An Introduction.* London: Verso, 1991.

Engstrom, Stephen P., and Jennifer Whiting. *Aristotle, Kant, and the Stoics: Rethinking Happiness and Duty.* Cambridge: Cambridge University Press, 1998.

Esposito, John L. *Islam: The Straight Path.* 3d ed. New York: Oxford University Press, 1998.

_____. *The Oxford History of Islam.* Oxford: Oxford University Press, 1999.

_____. *Unholy War: Terror in the Name of Islam.* Oxford: Oxford University Press, 2002.

Esposito, John L., and François Burgat. *Modernizing Islam: Religion in the Public Sphere in the Middle East and Europe.* New Brunswick, N.J.: Rutgers University Press, 2003.

Ewing, Alphonse B. *The USA Patriot Act.* New York: Novinka, 2002.

Fattah, Hassan M. *P2P: How Peer-to-Peer Technology Is Revolutionizing the Way We Do Business.* Chicago: Dearborn, 2002.

Feenberg, Andrew. *Critical Theory of Technology.* New York: Oxford University Press, 1991.

_____. *Questioning Technology.* London: Routledge, 1999.

Feenberg, Andrew, and Alastair Hannay. *Technology and the Politics of Knowledge.* Bloomington: Indiana University Press, 1995.

Florida, Richard L. *The Rise of the Creative Class: And How It's Transforming Work, Leisure, Community, and Everyday Life.* New York: Basic, 2002.

Foot, Michael, and Isaac Kramnick, eds. *Thomas Paine Reader.* Penguin Classics. Harmondsworth, U.K.: Penguin, 1987.

Foucault, Michel. *Discipline and Punish: The Birth of the Prison.* 2d ed. New York: Vintage, 1995.

_____. *The History of Sexuality.* New York: Vintage, 1988.

_____. *Madness and Civilization: A History of Insanity in the Age of Reason.* New York: Vintage, 1973.

_____. *The Order of Things: An Archaeology of the Human Sciences.* New York: Vintage, 1973.

Foucault, Michel, Graham Burchell, Colin Gordon, and Peter Miller. *The Foucault Effect: Studies in Governmentality: With Two Lectures by and an Interview with Michel Foucault.* Chicago: University of Chicago Press, 1991.

Foucault, Michel, and David Couzens Hoy. *Foucault: A Critical Reader.* Oxford: Blackwell, 1986.

Fox, Michael W. *Superpigs and Wondercorn: The Brave New World of Biotechnology and Where It All May Lead*. New York: Lyons & Burford, 1992.

Frank, Thomas. *The Conquest of Cool: Business Culture, Counterculture, and the Rise of Hip Consumerism*. Chicago: University of Chicago Press, 1997.

_____. *One Market Under God: Extreme Capitalism, Market Populism, and the End of Economic Democracy*. New York: Doubleday, 2000.

Frankel, Mark S., and Albert H. Teich. *The Genetic Frontier: Ethics, Law, and Policy*. Washington, D.C.: American Association for the Advancement of Science, 1994.

Friedman, Thomas L. *The Lexus and the Olive Tree*. New York: Anchor, 2000.

_____. *Longitudes and Attitudes: Exploring the World After September 11*. New York: Farrar Strauss & Giroux, 2002.

Galloway, Alexander R. *Protocol: How Control Exists After Decentralization*. Cambridge, Mass.: MIT Press, 2004.

Gans, Herbert J. *Democracy and the News*. Oxford: Oxford University Press, 2003.

_____. *Popular Culture and High Culture: An Analysis and Evaluation of Taste*. Rev. and updated ed. New York: Basic, 1999.

Garfinkel, Simson. *Database Nation: The Death of Privacy in the Twenty-first Century*. Cambridge, Mass.: O'Reilly, 2000.

_____. "Leaderless Resistance Today." *First Monday* 8, no. 3 (2003).

Geertz, Clifford. *Available Light: Anthropological Reflections on Philosophical Topics*. Princeton, N.J.: Princeton University Press, 2000.

_____. *The Interpretation of Cultures: Selected Essays*. New York: Basic, 1973.

Gilder, George F. *The Spirit of Enterprise*. New York: Simon & Schuster, 1984.

_____. *Telecosm: How Infinite Bandwidth Will Revolutionize Our World*. New York: Free Press, 2000.

Giroux, Henry A. *Public Spaces, Private Lives: Beyond the Culture of Cynicism*. Culture and Politics Series. Lanham, Md.: Rowman & Littlefield, 2001.

Gitlin, Todd. *Inside Prime Time*. New York: Pantheon, 1985.

_____. *Media Unlimited: How the Torrent of Images and Sounds Overwhelms Our Lives*. New York: Metropolitan, 2001.

_____. *The Sixties: Years of Hope, Days of Rage*. Toronto: Bantam, 1987.

_____. *The Twilight of Common Dreams: Why America Is Wracked by Culture Wars*. New York: Holt, 1996.

Glassgold, Peter. *Anarchy: An Anthology of Emma Goldman's Mother Earth*. Washington, D.C.: Counterpoint, 2001.

Glassner, Barry. *The Culture of Fear: Why Americans Are Afraid of the Wrong Things*. New York: Basic, 1999.

Godin, Seth. *Survival Is Not Enough: Zooming, Evolution, and the Future of Your Company*. New York: Free Press, 2002.

Goldfarb, Jeffrey C. *The Cynical Society: The Culture of Politics and the Politics of Culture in American Life*. Chicago: University of Chicago Press, 1991.

Goldman, Emma. *Red Emma Speaks: An Emma Goldman Reader*. Edited by Alix Kates Shulman. New York: Schocken, 1983.

Goldstein, Paul. *Changing the American Schoolbook: Law, Politics, and Technology*. Lexington Books Politics of Education Series. Lexington, Mass.: Lexington, 1978.

_____. *Copyright, Patent, Trademark, and Related State Doctrines: Cases and Materials on the Law of Intellectual Property*. Rev. 4th ed. University Casebook Series. Westbury, N.Y.: Foundation Press, 1999.

_____. *Copyright: Principles, Law, and Practice*. Boston: Little, Brown, 1989.

_____. *Copyright's Highway: From Gutenberg to the Celestial Jukebox*. New York: Hill & Wang, 1994.

_____. *International Copyright: Principles, Law, and Practice*. Oxford: Oxford University Press, 2001.

_____. *International Intellectual Property Law: Cases and Materials*. University Casebook Series. New York: Foundation Press, 2001.

Goldstein, Paul, Stella W. Lillick, and Ira S. Lillick. *Copyright*. 2d ed. Boston: Little, Brown, 1996.

Gould, Stephen Jay. *The Structure of Evolutionary Theory*. Cambridge, Mass.: Belknap Press of Harvard University Press, 2002.

Gourevitch, Philip. *We Wish to Inform You That Tomorrow We Will Be Killed with Our Families: Stories from Rwanda*. New York: Farrar Straus & Giroux, 1998.

Graham, Gordon. *The Internet: A Philosophical Inquiry*. London: Routledge, 1999.

Gramsci, Antonio, and David Forgacs. *The Gramsci Reader: Selected Writings, 1916–1935*. London: Lawrence & Wishart, 1999.

Granovetter, Mark. "Economic Action and Social Structure: The Problem of Embeddedness." *American Journal of Sociology* 91 (1985): 481–510.

Grossman, Wendy. *From Anarchy to Power: The Net Comes of Age*. New York: New York University Press, 2001.

Guerin, Daniel. *Anarchism: From Theory to Practice*. New York: Monthly Review Press, 1970.

Habermas, Jürgen. *Between Facts and Norms: Contributions to a Discourse Theory of Law and Democracy*. Studies in Contemporary German Social Thought. Cambridge, Mass.: MIT Press, 1996.

_____. *Knowledge and Human Interests*. Boston: Beacon, 1972.

_____. *The Philosophical Discourse of Modernity: Twelve Lectures*. Studies in Contemporary German Social Thought. Cambridge, Mass.: MIT Press, 1987.

_____. *The Structural Transformation of the Public Sphere: An Inquiry into a Category of Bourgeois Society*. Studies in Contemporary German Social Thought. Cambridge, Mass.: MIT Press, 1989.

_____. *The Theory of Communicative Action*. Boston: Beacon, 1984.

Habermas, Jürgen, and Maeve Cooke. *On the Pragmatics of Communication*. Studies in Contemporary German Social Thought. Cambridge, Mass.: MIT Press, 1998.

Hadas, Moses. *Essential Works of Stoicism*. Library of Basic Ideas. New York: Bantam, 1961.

Harari, Josué V. *Textual Strategies: Perspectives in Post-Structuralist Criticism*. Ithaca, N.Y.: Cornell University Press, 1979.

Hardt, Michael. *Gilles Deleuze: An Apprenticeship in Philosophy*. Minneapolis: University of Minnesota Press, 1993.

Hardt, Michael, and Antonio Negri. *Empire*. Cambridge, Mass.: Harvard University Press, 2000.

Hartley, John. *Popular Reality: Journalism, Modernity, Popular Culture*. London: Arnold, 1996.

_____. *Uses of Television*. Taylor & Francis e-Library ed. London: Routledge, 2002.

Hebdige, Dick. *Cut 'N' Mix: Culture, Identity, and Caribbean Music*. London: Methuen, 1987.

_____. *Subculture: The Meaning of Style*. New Accents. London: Routledge, 1991.

Heins, Marjorie. *Not in Front of the Children: "Indecency," Censorship, and the Innocence of Youth*. New York: Hill & Wang, 2001.

_____. *The Progress of Science and Useful Arts: Why Copyright Today Threatens Intellectual Freedom*. New York: Free Expression Policy Project, 2003.

_____. *Sex, Sin, and Blasphemy: A Guide to America's Censorship Wars*. New York: New Press/Norton, 1993.

Heins, Marjorie, and Christina Cho. *Internet Filters: A Public Policy Report*. New York: Free Expression Policy Project National Coalition Against Censorship, 2001.

Held, David, and Anthony G. McGrew. *The Global Transformations Reader: An Introduction to the Globalization Debate*. Cambridge: Polity, 2000.

Henriksen, M. A., A. Betz, M. V. Fuccillo, and J. E. Darnell Jr. "Negative Regulation of Stat92e by an N-Terminally Truncated Stat Protein Derived from an Alternative Promoter Site." *Genes Dev* 16, no. 18 (2002): 2379–2389.

Higgs, Eric S., Andrew Light, and David Strong. *Technology and the Good Life?* Chicago: University of Chicago Press, 2000.

Holm, Hans-Henrik, and Georg Sorensen. *Whose World Order? Uneven Globalization and the End of the Cold War*. Boulder, Colo.: Westview, 1995.

Honderich, Ted. *The Oxford Companion to Philosophy*. Oxford: Oxford University Press, 1995.

Honohan, Iseult. *Civic Republicanism*. Problems of Philosophy Series. New York: Routledge, 2002.

Howe, John R. *The Role of Ideology in the American Revolution*. American Problem Studies. New York: Holt, Rinehart & Winston, 1970.

Innis, Harold Adams, and Marshall McLuhan. *The Bias of Communication*. Toronto: University of Toronto Press, 1964.

Irwin, William. *The Matrix and Philosophy: Welcome to the Desert of the Real*. Vol. 3 of *Popular Culture and Philosophy*. Chicago: Open Court, 2002.

Israel, Jonathan Irvine. *Radical Enlightenment: Philosophy and the Making of Modernity, 1650–1750*. Oxford: Oxford University Press, 2001.

Jacker, Corinne. *The Black Flag of Anarchy: Antistatism in the United States*. New York: Scribner, 1968.

Jacobs, Jane. *Cities and the Wealth of Nations: Principles of Economic Life*. New York: Random House, 1984.

_____. *The Death and Life of Great American Cities*. New York: Modern Library, 1993.

_____. *The Nature of Economies*. New York: Modern Library, 2000.

_____. *Systems of Survival: A Dialogue on the Moral Foundations of Commerce and Politics*. New York: Random House, 1992.

Jacobson, Matthew Frye. *Barbarian Virtues: The United States Encounters Foreign Peoples at Home and Abroad, 1876–1917*. New York: Hill & Wang, 2000.

Jefferson, Thomas, and William Harwood Peden. *Notes on the State of Virginia*. Chapel Hill: Institute of Early American History and Culture/University of North Carolina Press, 1955.

Jenkins, Henry. *Textual Poachers: Television Fans and Participatory Culture*. Studies in Culture and Communication. New York: Routledge, 1992.

_____. *"What Made Pistachio Nuts?" Anarchistic Comedy and the Vaudeville Aesthetic*. New York: Columbia University Press, 1993.

Jenkins, Philip. *Beyond Tolerance: Child Pornography on the Internet*. New York: New York University Press, 2001.

_____. *Intimate Enemies: Moral Panics in Contemporary Great Britain*. Social Problems and Social Issues. New York: Aldine de Gruyter, 1992.

_____. *Moral Panic: Changing Concepts of the Child Molester in Modern America*. New Haven, Conn.: Yale University Press, 1998.

_____. *The Next Christendom: The Rise of Global Christianity*. Oxford: Oxford University Press, 2002.

_____. *Pedophiles and Priests: Anatomy of a Contemporary Crisis*. New York: Oxford University Press, 1996.

_____. *Synthetic Panics: The Symbolic Politics of Designer Drugs*. New York: New York University Press, 1999.

Kalathil, Shanthi, and Taylor C. Boas. *Open Networks, Closed Regimes: The Impact of the Internet on Authoritarian Regimes*. Washington, D.C.: Carnegie Endowment for International Peace, 2003.

Kaplan, Robert D. "The Coming Anarchy." *Atlantic Monthly*, February 1994.

_____. *The Coming Anarchy: Shattering the Dreams of the Post Cold War*. New York: Random House, 2000.

Karabell, Zachary. *What's College For? The Struggle to Define American Higher Education*. New York: Basic, 1998.

Katz, Adam. *Postmodernism and the Politics of "Culture."* Cultural Studies. Boulder, Colo.: Westview, 2000.

Keller, Evelyn Fox. *The Century of the Gene*. Cambridge, Mass.: Harvard University Press, 2000.

_____. *Reflections on Gender and Science*. New Haven, Conn.: Yale University Press, 1985.

Kelly, Kevin. *New Rules for the New Economy: Ten Radical Strategies for a Connected World*. New York: Viking, 1998.

_____. *Out of Control: The Rise of Neo-Biological Civilization*. Reading, Mass.: Addison-Wesley, 1994.

Kepel, Gilles. *Jihad: The Trail of Political Islam*. Cambridge, Mass.: Belknap Press of Harvard University Press, 2002.

Kevles, Daniel J., and Leroy E. Hood. *The Code of Codes: Scientific and Social Issues in the Human Genome Project*. Cambridge, Mass.: Harvard University Press, 1992.

Khalilzad, Zalmay M., and John P. White. *Strategic Appraisal: The Changing Role of Information in Warfare*. Santa Monica, Calif.: Rand, 1999.

Klein, Naomi. *No Space, No Choice, No Jobs, No Logo: Taking Aim at the Brand Bullies*. New York: Picador USA, 2000.

Klein, Naomi, and Debra Ann Levy. *Fences and Windows: Dispatches from the Front Lines of the Globalization Debate*. New York: Picador USA, 2002.

Kloet, Jeroen de. "Digitisation and Its Asian Discontents: The Internet, Politics, and Hacking in China and Indonesia." *First Monday* 7, no. 9 (2002).

Knezo, Genevieve Johanna. *Federal Security Controls on Scientific and Technical Information: History and Current Controversy*. New York: Nova Science Publishers, 2003.

Kranich, Nancy C. *Libraries and Democracy: The Cornerstones of Liberty*. Chicago: American Library Association, 2001.

Krebs, Viola. "The Impact of the Internet on Myanmar." *First Monday* 6, no. 5 (2001).

Kropotkin, Petr Alekseevich. *Anarchism and Anarchist Communism: Two Essays*. Edited by Nicolas Walter. Anarchist Classics. London: Freedom Press, 1987.

_____. *Mutual Aid: A Factor of Evolution*. 1902. London: Allen Lane, 1972.

Larkin, Brian. "Bollywood Comes to Nigeria." *Samar* 8 (1997).

_____. "Degrading Images, Distorted Sounds: Nigerian Video and the Infrastructure of Piracy." *Public Culture* 16, no. 3 (2003).

Leon-Garcia, Alberto, and Indra Widjaja. *Communication Networks: Fundamental Concepts and Key Architectures*. McGraw-Hill Series in Computer Science. Boston: McGraw-Hill, 2000.

Lessig, Lawrence. *Code and Other Laws of Cyberspace*. New York: Basic, 1999.

_____. *The Future of Ideas: The Fate of the Commons in a Connected World*. New York: Random House, 2001.

Lester, J. C. *Escape from Leviathan: Liberty, Welfare, and Anarchy Reconciled*. New York: St. Martin's, 2000.

Levine, Lawrence W. *Black Culture and Black Consciousness: Afro-American Folk Thought from Slavery to Freedom*. New York: Oxford University Press, 1977.

Lévy, Pierre. *Becoming Virtual: Reality in the Digital Age*. New York: Plenum Trade, 1998.

_____. *Collective Intelligence: Mankind's Emerging World in Cyberspace*. Helix Books. Cambridge, Mass.: Perseus, 1999.

_____. *Cyberculture*. Vol. 4 of *Electronic Mediations*. Minneapolis: University of Minnesota Press, 2001.

Lewis, Justin, and Toby Miller. *Critical Cultural Policy Studies: A Reader*. Oxford: Blackwell, 2002.

Lewis, Michael. *Next: The Future Just Happened*. New York: Norton, 2001.

Library Quarterly, January 1931–.

Liebowitz, S. J. *Re-Thinking the Network Economy: The True Forces That Drive the Digital Marketplace*. New York: Amacom, 2002.

Liebowitz, S. J., and Stephen Margolis. *Winners, Losers, and Microsoft: Competition and Antitrust in High Technology*. Rev. ed. Independent Studies in Political Economy. Oakland, Calif.: Independent Institute, 2001.

Light, Andrew. *Reel Arguments: Film, Philosophy, and Social Criticism*. Boulder, Colo.: Westview, 2003.

Light, Andrew, and Jonathan M. Smith. *The Production of Public Space*. Vol. 2 of *Philosophy and Geography*. Lanham, Md.: Rowman & Littlefield, 1998.

Lippmann, Walter. *The Phantom Public*. New York: Harcourt Brace, 1925.

_____. *Public Opinion*. New York: Harcourt Brace, 1922.

Litman, Jessica. *Digital Copyright: Protecting Intellectual Property on the Internet*. Amherst, N.Y.: Prometheus, 2001.

Loewenstein, Werner R. *The Touchstone of Life: Molecular Information, Cell Communication, and the Foundations of Life*. New York: Oxford University Press, 1999.

Ludlow, Peter. *Crypto Anarchy, Cyberstates, and Pirate Utopias*. Digital Communication. Cambridge, Mass.: MIT Press, 2001.

_____. *High Noon on the Electronic Frontier: Conceptual Issues in Cyberspace*. Digital Communication. Cambridge, Mass.: MIT Press, 1996.

Luttwak, Edward. *Turbo-Capitalism: Winners and Losers in the Global Economy*. New York: HarperCollins, 1999.

Magder, Ted. *Canada's Hollywood: The Canadian State and Feature Films*. The State and Economic Life, no. 18. Toronto: University of Toronto Press, 1993.

_____. *Franchising the Candy Store: Split-Run Magazines and a New International Regime for Trade in Culture*. Canadian-American Public Policy, no. 34. Orono, Me.: Canadian-American Center University of Maine, 1998.

Mak, Geert. *Amsterdam: A Brief Life of the City*. London: Harvill, 1999.

Mandel, Michael J. *The Coming Internet Depression: Why the High-Tech Boom Will Go Bust, Why the Crash Will Be Worse Than You Think, and How to Prosper Afterwards*. New York: Basic, 2000.

Manuel, Peter Lamarche. *Cassette Culture: Popular Music and Technology in North India*. Chicago Studies in Ethnomusicology. Chicago: University of Chicago Press, 1993.

Maskus, Keith E. *Intellectual Property Rights in the Global Economy*. Washington, D.C.: Institute for International Economics, 2000.

May, Christopher. "Digital Rights Management and the Breakdown of Social Norms." *First Monday* 8, no. 11 (2003).

_____. *A Global Political Economy of Intellectual Property Rights: The New Enclosures?* Routledge/Ripe Studies in Global Political Economy. London: Routledge, 2000.

_____. *The Information Society: A Sceptical View*. Cambridge: Polity, 2002.

McLean, George N. *The Rise and Fall of Anarchy in America: From Its Incipient Stage to the First Bomb Thrown in Chicago. A Comprehensive Account of the Great Conspiracy Culminating in the Haymarket Massacre, May 4th, 1886 . . . The Apprehension, Trial, Conviction and Execution of the Leading Conspirators*. Chicago, Philadelphia: R. G. Badoux, 1888.

McLuhan, Marshall. *Understanding Media: The Extensions of Man*. Cambridge, Mass.: MIT Press, 1994.

McLuhan, Marshall, and Bruce R. Powers. *The Global Village: Transformations in World Life and Media in the Twenty-first Century*. New York: Oxford University Press, 1989.

McNeill, Don. *Moving Through Here*. New York: Knopf, 1970.

Micklethwait, John, and Adrian Wooldridge. *A Future Perfect: The Challenge and Hidden Promise of Globalization*. New York: Times Books, 2000.

Mill, John Stuart. *On Liberty*. Edited by David Spitz. Norton Critical Edition. New York: Norton, 1975.

Miller, Jim. *"Democracy Is in the Streets."* New York: Simon & Schuster, 1987.

_____. *The Passion of Michel Foucault*. New York: Simon & Schuster, 1993.

Miller, Toby. *Global Hollywood*. London: British Film Institute, 2001.

_____. *Technologies of Truth: Cultural Citizenship and the Popular Media*. Vol. 2 of *Visible Evidence*. Minneapolis: University of Minnesota Press, 1998.

_____. *The Well-Tempered Self: Citizenship, Culture, and the Postmodern Subject*. Baltimore: Johns Hopkins University Press, 1993.

Miller, Toby, and George Yudice. *Cultural Policy*. London: Sage, 2002.

Mills, C. Wright. *The Power Elite*. Oxford: Oxford University Press, 2000.

_____. *The Sociological Imagination*. Oxford: Oxford University Press, 2000.

_____. *Sociology and Pragmatism: The Higher Learning in America*. New York: Paine-Whitman, 1964.

Milone, Mark G. "Hactivism: Securing the National Infrastructure." *Business Lawyer* 58 (2002).

Mitten, Christopher. *Shawn Fanning: Napster and the Music Revolution*. Techies. Brookfield, Conn.: Twenty-First Century Books, 2002.

Money Laundering, Terrorism, and Financial Institutions: Law, Regulation, Compliance, USA Patriot Act Monitor. Kingston, N.J.: Civic Research Institute, 2002.

Moore, Mike. *A World Without Walls: Freedom, Development, Free Trade, and Global Governance.* Cambridge: Cambridge University Press, 2003.

Mosco, Vincent. *The Pay-Per Society: Computers and Communication in the Information Age: Essays in Critical Theory and Public Policy.* Toronto: Garamond, 1989.

———. *The Political Economy of Communication: Rethinking and Renewal.* Media, Culture, and Society Series. London: Sage, 1996.

Mosco, Vincent, and Dan Schiller. *Continental Order? Integrating North America for Cybercapitalism.* Critical Media Studies. Lanham, Md.: Rowman & Littlefield, 2001.

Mosco, Vincent, and Janet Wasko. *The Political Economy of Information.* Studies in Communication and Society. Madison: University of Wisconsin Press, 1988.

Mueller, Milton. *Ruling the Root: Internet Governance and the Taming of Cyberspace.* Cambridge, Mass.: MIT Press, 2002.

Mueller, Milton, and Zixiang Tan. *China in the Information Age: Telecommunications and the Dilemmas of Reform.* Washington Papers, no. 169. Westport, Conn.: Praeger, 1997.

Mumford, Lewis. *Art and Technics.* Bampton Lectures in America, no. 4. New York: Columbia University Press, 1952.

———. *Technics and Civilization.* New York: Harcourt Brace, 1963.

Navia, Luis E. *Classical Cynicism: A Critical Study.* Contributions in Philosophy, no. 58. Westport, Conn.: Greenwood, 1996.

———. *Diogenes of Sinope: The Man in the Tub.* Contributions in Philosophy, no. 67. Westport, Conn.: Greenwood, 1998.

———. *The Philosophy of Cynicism: An Annotated Bibliography.* Bibliographies and Indexes in Philosophy, no. 4. Westport, Conn.: Greenwood, 1995.

Negroponte, Nicholas. *Being Digital.* New York: Knopf, 1995.

Nehring, Neil. *Flowers in the Dustbin: Culture, Anarchy, and Postwar England.* Ann Arbor: University of Michigan Press, 1993.

Nozick, Robert. *Anarchy, State, and Utopia.* New York: Basic, 1974.

Nozick, Robert, and Jeffrey Paul. *Reading Nozick: Essays on Anarchy, State, and Utopia.* Philosophy and Society. Totowa, N.J.: Rowman & Littlefield, 1981.

Nussbaum, Martha Craven. *Cultivating Humanity: A Classical Defense of Reform in Liberal Education.* Cambridge, Mass.: Harvard University Press, 1997.

———. *The Fragility of Goodness: Luck and Ethics in Greek Tragedy and Philosophy.* Rev. ed. Cambridge: Cambridge University Press, 2001.

Nussbaum, Martha Craven, and Joshua Cohen. *For Love of Country?* New Democracy Forum. Boston: Beacon, 2002.

———. *For Love of Country: Debating the Limits of Patriotism.* Boston: Beacon, 1996.

Nussbaum, Martha Craven, and Angela Kallhoff. *Martha C. Nussbaum: Ethics and Political Philosophy: Lecture and Colloquium in Münster 2000.* Vol. 4 of *Münsteraner Vorlesungen zur Philosophie.* Münster: Lit Verlag, 2001.

Oldfield, Adrian. *Citizenship and Community: Civic Republicanism and the Modern World*. London: Routledge, 1990.

Oram, Andrew. *Peer-to-Peer: Harnessing the Power of Disruptive Technologies*. Cambridge, Mass.: O'Reilly, 2001.

Orwell, George, and Lionel Trilling. *Homage to Catalonia*. San Diego: Harcourt Brace, 1980.

Owen, G. E. L., and Martha Craven Nussbaum. *Logic, Science, and Dialectic: Collected Papers in Greek Philosophy*. Ithaca, N.Y.: Cornell University Press, 1986.

Pagels, Elaine H. *Beyond Belief: The Secret Gospel of Thomas*. New York: Random House, 2003.

Paine, Thomas. *The Age of Reason. Part the Second. Being an Investigation of True and Fabulous Theology*. London: H. D. Symonds, 1795.

———. *Rights of Man*. Everyman's Library: Philosophy and Theology, no. 718. London: J. M. Dent, 1935.

———. *Rights of Man, Common Sense, and Other Political Writings*. Edited by Mark Philp. World's Classics. Oxford: Oxford University Press, 1995.

Patterson, L. Ray. *Copyright in Historical Perspective*. Nashville, Tenn.: Vanderbilt University Press, 1968.

Patterson, L. Ray, and Stanley W. Lindberg. *The Nature of Copyright: A Law of Users' Rights*. Athens: University of Georgia Press, 1991.

Patterson, Orlando. *Freedom in the Making of Western Culture*. Edited by Orlando Patterson. Vol. 1 of *Freedom*. New York: Basic, 1991.

———. *Slavery and Social Death: A Comparative Study*. Cambridge, Mass.: Harvard University Press, 1982.

Perelman, Michael. *Steal This Idea: Intellectual Property Rights and the Corporate Confiscation of Creativity*. New York: Palgrave, 2002.

Pink, Daniel H. *Free Agent Nation: How America's New Independent Workers Are Transforming the Way We Live*. New York: Warner, 2001.

Politi, Alessandro. "The Citizen as Intelligence Minuteman." *International Journal of Intelligence and Counterintelligence* 16 (2003): 34–38.

Popper, Karl Raimund. *All Life Is Problem Solving*. London: Routledge, 1999.

———. *The Logic of Scientific Discovery*. London: Routledge, 1992.

———. *The Open Society and Its Enemies*. London: Routledge, 1945.

———. *The Poverty of Historicism*. Boston: Beacon, 1957.

Popper, Karl Raimund, and Mark Amadeus Notturno. *The Myth of the Framework: In Defence of Science and Rationality*. London: Routledge, 1994.

Postman, Neil. *Amusing Ourselves to Death: Public Discourse in the Age of Show Business*. New York: Penguin, 1986.

———. *Building a Bridge to the Eighteenth Century: How the Past Can Improve Our Future*. New York: Knopf/Random House, 1999.

———. *Technopoly: The Surrender of Culture to Technology*. New York: Vintage, 1993.

Postrel, Virginia I. *The Future and Its Enemies: The Growing Conflict over Creativity, Enterprise, and Progress.* New York: Free Press, 1998.

Press, Eyal, and Jennifer Washburn. "The Kept University." *Atlantic Monthly*, March 2000.

Price, Monroe Edwin. *Media and Sovereignty: The Global Information Revolution and Its Challenge to State Power.* Cambridge, Mass.: MIT Press, 2002.

_____. *Television, the Public Sphere, and National Identity.* Oxford: Clarendon, 1995.

Price, Monroe Edwin, and Melville B. Nimmer. *Resuscitating a Collaboration with Melville Nimmer: Moral Rights and Beyond.* Occasional Papers in Intellectual Property from Benjamin N. Cardozo School of Law, Yeshiva University no. 3. New York: Benjamin N. Cardozo School of Law, Yeshiva University, 1998.

Price, Monroe Edwin, Roger G. Noll, and Lloyd Morrisett. *A Communications Cornucopia: Markle Foundation Essays on Information Policy.* Washington, D.C.: Brookings Institution Press, 1998.

Price, Monroe Edwin, Beata Rozumilowicz, and Stefaan Verhulst. *Media Reform: Democratizing the Media, Democratizing the State.* Routledge Research in Cultural and Media Studies. London: Routledge, 2002.

Price, Monroe Edwin, and Stefaan Verhulst. *Broadcasting Reform in India: Media Law from a Global Perspective.* Law in India Series. Delhi: Oxford University Press, 1998.

Randall, Alice. *The Wind Done Gone.* Boston: Houghton Mifflin, 2001.

Rankin, H. D. *Sophists, Socratics, and Cynics.* London: Croon Helm, 1983.

Rauchway, Eric. *Murdering Mckinley: The Making of Theodore Roosevelt's America.* New York: Hill & Wang, 2003.

Reams, Bernard D., and Christopher Anglim. *USA Patriot Act: A Legislative History of the Uniting and Strengthening of America by Providing Appropriate Tools Required to Intercept and Obstruct Terrorism Act, Public Law No. 107–56 (2001).* 5 vols. Buffalo, N.Y.: Hein, 2002.

Reed, Sally Gardner. *Creating the Future: Essays on Librarianship in an Age of Great Change.* Jefferson, N.C.: McFarland, 1996.

Regulatory Compliance Associates and Sheshunoff Information Services. *Special Report, USA Patriot Act: A Guide to Regulatory Compliance.* Austin, Tex.: Thomson/Sheshunoff, 2002.

Reiman, Donald H., and Mary A. Quinn. *Percy Bysshe Shelley.* Manuscripts of the Younger Romantics. New York: Garland, 1985.

Rheingold, Howard. *Smart Mobs: The Next Social Revolution.* Cambridge, Mass.: Perseus, 2002.

_____. *The Virtual Community: Homesteading on the Electronic Frontier.* Cambridge, Mass.: MIT Press, 2000.

Riesman, David, Nathan Glazer, and Reuel Denney. *The Lonely Crowd: A Study of the Changing American Character.* Yale Nota Bene. New Haven, Conn.: Yale University Press, 2001.

Riis, Jacob A. *How the Other Half Lives: Studies Among the Tenements of New York.* Edited by David Leviatin. Bedford Series in History and Culture. Boston: Bedford Books of St. Martin's Press, 1996.

Rimmer, Kathy Bowrey, and Matthew Rimmer. "Rip, Mix, Burn: The Politics of Peer to Peer and Copyright Law." *First Monday* 7, no. 8 (2002).

Ronfeldt, David F., John Arquilla, Graham E. Fuller, Melissa Fuller. *The Zapatista "Social Netwar" in Mexico.* Washington, D.C.: RAND Corporation, 1998.

Rose, Tricia. *Black Noise: Rap Music and Black Culture in Contemporary America.* Music/Culture. Hanover, N.H.: Wesleyan University Press/University Press of New England, 1994.

Rosecrance, Richard N. *The Rise of the Virtual State: Wealth and Power in the Coming Century.* New York: Basic, 1999.

Rosen, Jonathan. *The Talmud and the Internet: A Journey Between Worlds.* New York: Farrar Straus & Giroux, 2000.

Ross, Andrew, and Tricia Rose. *Microphone Fiends: Youth Music and Youth Culture.* New York: Routledge, 1994.

Rossinow, Douglas C. *The Politics of Authenticity: Liberalism, Christianity, and the New Left in America.* Contemporary American History Series. New York: Columbia University Press, 1998.

Rothfield, Lawrence. *Unsettling "Sensation": Arts-Policy Lessons from the Brooklyn Museum of Art Controversy.* Rutgers Series on the Public Life of the Arts. New Brunswick, N.J.: Rutgers University Press, 2001.

Roussopoulos, Dimitrios I. *The Anarchist Papers.* Montreal: Black Rose, 1986.

_____. *The Anarchist Papers.* Vol. 2. Montreal: Black Rose, 1989.

_____. *The Anarchist Papers.* Vol. 3. Montreal: Black Rose, 1990.

_____. *The Public Place: Citizen Participation in the Neighbourhood and the City.* Montreal: Black Rose, 1999.

Ryan, Michael P. *Knowledge Diplomacy: Global Competition and the Politics of Intellectual Property.* Washington, D.C.: Brookings Institution Press, 1998.

Ryan, Nick. *Homeland: Into a World of Hate.* Edinburgh: Mainstream, 2003.

Sabin, Roger. *Punk Rock: So What? The Cultural Legacy of Punk.* London: Routledge, 1999.

Samuels, Edward B. *The Illustrated Story of Copyright.* New York: Thomas Dunne/St. Martin's, 2000.

Sassen, Saskia. *Cities in a World Economy.* 2d ed. Sociology for a New Century. Thousand Oaks, Calif.: Pine Forge, 2000.

_____. *Globalization and Its Discontents: Essays on the New Mobility of People and Money.* New York: New Press, 1998.

_____. *Global Networks, Linked Cities.* New York: Routledge, 2002.

_____. "The Internet and Sovereignty." *Indiana Journal of Global Legal Studies,* Special Issue, Winter 1999.

Sayre, Farrand. *The Greek Cynics.* Baltimore: J. H. Furst, 1948.

Schaack, Michael J. *Anarchy and Anarchists: Anti-Movements in America*. New York: Arno, 1977.

Schechter, Danny. *Falun Gong's Challenge to China: Spiritual Practice or "Evil Cult"? A Report and Reader*. New York: Akashic, 2000.

Schiller, Dan. *Digital Capitalism Networking the Global Market System*. Cambridge, Mass.: MIT Press, 1999.

_____. *Theorizing Communication: A History*. New York: Oxford University Press, 1996.

Schiller, Herbert I. *Culture, Inc.: The Corporate Takeover of Public Expression*. New York: Oxford University Press, 1989.

_____. *Information and the Crisis Economy, Communication, and Information Science*. Norwood, N.J.: Ablex, 1984.

_____. *Information Inequality: The Deepening Social Crisis in America*. New York: Routledge, 1996.

Schneier, Bruce. *Applied Cryptography: Protocols, Algorithms, and Source Code in C*. 2d ed. New York: Wiley, 1996.

_____. *Secrets and Lies: Digital Security in a Networked World*. New York: Wiley, 2000.

Schneier, Bruce, and David Banisar. *The Electronic Privacy Papers: Documents on the Battle for Privacy in the Age of Surveillance*. New York: Wiley, 1997.

Sell, Susan K. *Power and Ideas: North-South Politics of Intellectual Property and Antitrust*. SUNY Series in Global Politics. Albany: State University of New York Press, 1998.

_____. *Private Power, Public Law: The Globalization of Intellectual Property Rights*. Cambridge Studies in International Relations, no. 88. Cambridge: Cambridge University Press, 2003.

Sen, Amartya Kumar. *Development as Freedom*. New York: Knopf, 1999.

_____. *Rationality and Freedom*. Cambridge, Mass.: Belknap Press of Harvard University Press, 2002.

Sheehan, Sean M. *Anarchism: Focus on Contemporary Issues*. London: Reaktion, 2003.

Shelley, Percy Bysshe. *Selected Poetry and Prose*. Edited by Kenneth Neill Cameron. Rinehart Editions, no. 49. New York: Rinehart, 1951.

Shulman, Seth. *Owning the Future*. Boston: Houghton Mifflin, 1999.

_____. *Unlocking the Sky: Glenn Hammond Curtiss and the Race to Invent the Airplane*. New York: HarperCollins, 2002.

Shusterman, Richard. *Bourdieu: A Critical Reader*. Oxford: Blackwell, 1999.

_____. *Practicing Philosophy: Pragmatism and the Philosophical Life*. New York: Routledge, 1997.

_____. *Pragmatist Aesthetics: Living Beauty, Rethinking Art*. Cambridge, Mass.: Blackwell, 1992.

Sloterdijk, Peter. *Critique of Cynical Reason*. Vol. 40 of *Theory and History of Literature*. Minneapolis: University of Minnesota Press, 1987.

Small, Christopher. *Music of the Common Tongue: Survival and Celebration in African American Music*. Music/Culture. Hanover, N.H.: University Press of New England, 1998.

Smith, Rogers M. *Civic Ideals: Conflicting Visions of Citizenship in U.S. History*. Yale ISPS Series. New Haven, Conn.: Yale University Press, 1997.

Sober, Elliott. *Conceptual Issues in Evolutionary Biology: An Anthology*. Cambridge, Mass.: MIT Press, 1984.

Solidarity. *Confronting Globalization: The Battle of Seattle and Beyond*. Detroit: Solidarity, 2000.

Southern, Eileen. *The Music of Black Americans: A History*. 2d ed. New York: Norton, 1983.

――――. *Readings in Black American Music*. 2d ed. New York: Norton, 1983.

Spar, Debora L. *Ruling the Waves: Cycles of Discovery, Chaos, and Wealth from the Compass to the Internet*. New York: Harcourt, 2001.

Sprigman, Chris. Hacking for Free Speech. Available at http://practice.findlaw.com/hack–0703.html.

Steel, Ronald. *Walter Lippmann and the American Century*. Boston: Little, Brown, 1980.

Stefik, Mark. *The Internet Edge: Social, Legal, and Technological Challenges for a Networked World*. Cambridge, Mass.: MIT Press, 1999.

Stiglitz, Joseph E. *Economics*. 2d ed. New York: Norton, 1997.

――――. *Globalization and Its Discontents*. New York: Norton, 2002.

――――. *Information and Capital Markets*. National Bureau of Economic Research Working Paper, no. 678. Cambridge, Mass.: National Bureau of Economic Research, 1981.

Stober, Dan, and Ian Hoffman. *A Convenient Spy: Wen Ho Lee and the Politics of Nuclear Espionage*. New York: Simon & Schuster, 2001.

Sulston, John, and Georgina Ferry. *The Common Thread: A Story of Science, Politics, Ethics, and the Human Genome*. Washington, D.C.: Joseph Henry, 2002.

Sunstein, Cass R. *Constitutional Myth-Making: Lessons from the Dred Scott Case*. Occasional Papers from the Law School, the University of Chicago, no. 37. Law School of the University of Chicago, 1996.

――――. *Democracy and the Problem of Free Speech*. New York: Free Press, 1993.

――――. *Designing Democracy: What Constitutions Do*. Oxford: Oxford University Press, 2001.

――――. *Free Markets and Social Justice*. New York: Oxford University Press, 1997.

――――. *Republic.Com*. Princeton, N.J.: Princeton University Press, 2001.

Taylor, Mark C. *The Moment of Complexity: Emerging Network Culture*. Chicago: University of Chicago Press, 2001.

Thomas, Timothy L. "Al Qaeda and the Internet: The Danger of 'Cyberplanning.'" *Parameters: U.S. Army War College Quarterly* (2003): 112–123.

Tocqueville, Alexis de. *Democracy in America.* Translated, edited, and with an introduction by Harvey C. Mansfield and Delba Winthrop. Chicago: University of Chicago Press, 2000.

Tolstoy, Leo. *Some Social Remedies: Socialism, Anarchy; Henry Georgism and the Land Question, Communism, Etc. Collected from the Recent and Unpublished Writings of Leo Tolstoy.* Maldon, U.K.: Essex, 1900.

Trend, David. *Cultural Democracy: Politics, Media, New Technology.* SUNY Series, Interruptions—Border Testimony(ies) and Critical Discourse/S. Albany: State University of New York Press, 1997.

Trilling, Lionel. *Matthew Arnold.* New York. Norton, 1939

Tulloch, John, and Henry Jenkins. *Science Fiction Audiences: Watching Doctor Who and Star Trek.* Popular Fiction Series. London: Routledge, 1995.

Turkle, Sherry. *Life on the Screen: Identity in the Age of the Internet.* New York: Simon & Schuster, 1997.

_____. *The Second Self: Computers and the Human Spirit.* New York: Simon & Schuster, 1984.

Turner, Victor Witter. *The Forest of Symbols: Aspects of Ndembu Ritual.* Ithaca, N.Y.: Cornell University Press, 1967.

Turner, Victor Witter, and Edward M. Bruner. *The Anthropology of Experience.* Urbana: University of Illinois Press, 1986.

United States House of Representatives. Committee on Financial Services. *Terrorist Financing: Implementation of the USA Patriot Act: Hearing Before the Committee on Financial Services, U.S. House of Representatives, 107th Cong., 2d sess., September 19, 2002.* Washington, D.C.: GPO, 2002.

United States House of Representatives. Committee on Government Reform. Special Investigations Division. *Children's Access to Pornography Through Internet File-Sharing Programs Prepared for Rep. Henry A. Waxman and Rep. Steve Largent.* Washington, D.C.: The House of Representatives, 2001.

United States Senate. Committee on Banking Housing and Urban Affairs. *The Financial War on Terrorism and the Administration's Implementation of Title III of the USA Patriot Act: Hearing Before the Committee on Banking, Housing, and Urban Affairs, United States Senate, 107th Cong., 2d sess., on the Administration's Implementation of the Anti-Money Laundering Provisions (Title III) of the USA Patriot Act (Public Law 107–56), and Its Efforts to Disrupt Terrorist Financing Activities, January 29, 2002.* Washington, D.C.: GPO, 2003.

United States Senate. Committee on the Judiciary. *Utah's Digital Economy and the Future: Peer-to-Peer and Other Emerging Technologies: Hearing Before the Committee on the Judiciary, United States Senate, 106th Cong., 2d sess., October 9, 2000, Provo, Ut, S. Hrg., 106–1070.* Washington, D.C.: GPO, 2001.

Uniting and Strengthening America by Providing Appropriate Tools Required to Intercept and Obstruct Terrorism (USA Patriot Act) Act of 2001. 107th Congress, 1st sess., H.R. 3162.

Vaidhyanathan, Siva. *Copyrights and Copywrongs: The Rise of Intellectual Property and How It Threatens Creativity.* New York: New York University Press, 2001.

_____. "Cultural Policy and the Art of Commerce." *Chronicle of Higher Education,* June 22, 2001, B7.

_____. "Unoriginal Sins: Copyright and American Culture." PhD diss., University of Texas, 1999.

Vaidhyanathan, Vishnampet S. *Regulation and Control Mechanisms in Biological Systems.* Prentice Hall Biophysics and Bioengineering Series. Englewood Cliffs, N.J.: PTR Prentice Hall, 1993.

Veblen, Thorstein. *The Theory of the Leisure Class: An Economic Study of Institutions.* With an Introduction by C. Wright Mills. Mentor Book. New York: New American Library, 1959.

Vegh, Sandor. "Hacktivists or Cyberterrorists? The Changing Media Discourse on Hacking." *First Monday* 7, no. 10 (2002).

Viroli, Maurizio. *For Love of Country: An Essay on Patriotism and Nationalism.* New York: Clarendon, 1995.

_____. *Jean-Jacques Rousseau and the "Well-Ordered Society."* Cambridge: Cambridge University Press, 1988.

_____. *Machiavelli.* Founders of Modern Political and Social Thought. Oxford: Oxford University Press, 1998.

_____. *Niccolò's Smile: A Biography of Machiavelli.* New York: Farrar Straus & Giroux, 2000.

_____. *Republicanism.* New York: Hill & Wang, 2002.

Warner, Michael. *The Letters of the Republic: Publication and the Public Sphere in Eighteenth-Century America.* Cambridge, Mass.: Harvard University Press, 1990.

_____. *Publics and Counterpublics.* New York: Zone/MIT Press, 2002.

Watts, Duncan J. *Six Degrees: The Science of a Connected Age.* New York: Norton, 2003.

_____. *Small Worlds: The Dynamics of Networks Between Order and Randomness.* Princeton Studies in Complexity. Princeton, N.J.: Princeton University Press, 1999.

Weber, Max. *Basic Concepts in Sociology.* 6th ed. New York: Citadel, 1969.

Wellman, Barry. *Networks in the Global Village: Life in Contemporary Communities.* Boulder, Colo.: Westview, 1999.

Whimster, Sam. *Max Weber and the Culture of Anarchy.* New York: St. Martin's, 1999.

Widner, Jennifer. "States and Statelessness in Late Twentieth-Century Africa." *Daedalus,* 1995, pp. 129–154.

Willard, Charles Arthur. *Liberalism and the Problem of Knowledge: A New Rhetoric for Modern Democracy, New Practices of Inquiry.* Chicago: University of Chicago Press, 1996.

Wishart, Adam, and Regula Bochsler. *Leaving Reality Behind: The Battle for the Soul of the Internet.* London: Fourth Estate, 2002.

Wright, Robert. *Nonzero: The Logic of Human Destiny*. New York: Vintage, 2001.

Yang, E., M. A. Henriksen, O. Schaefer, N. Zakharova, and J. E. Darnell Jr. "Dissociation Time from DNA Determines Transcriptional Function in a Stat1 Linker Mutant." *J Biol Chem* 277, no. 16 (2002): 13455–13462.

Yuen, Eddie, Daniel Burton-Rose, and George N. Katsiaficas. *The Battle of Seattle: The New Challenge to Capitalist Globalization*. New York: Soft Skull, 2002.

Zuidervaart, Lambert, and Henry Luttikhuizen. *The Arts, Community, and Cultural Democracy: Cross-Currents in Religion and Culture*. New York: St. Martin's, 2000.

Zweiger, Gary. *Transducing the Genome: Information, Anarchy, and Revolution in the Biomedical Sciences*. New York: McGraw-Hill, 2001.

INDEX

CPSIA information can be obtained at www.ICGtesting.com
Printed in the USA
BVOW01s0040110414

350331BV00001B/55/P